Charles Kovacs

Botany

Waldorf Education Resources

First published in volume form in 2005

© 2005 Estate of Charles Kovacs
Fifth printing 2017

The author has asserted his right under the
Copyright, Design and Patents Act 1988
to be identified as the Author of this Work

All right reserved. No part of this publication may
be reproduced without the prior permission of
Floris Books, Edinburgh

www.florisbooks.co.uk

British Library CIP Data available
ISBN 978-086315-537-6
Printed in Great Britain
by Bell & Bain, Ltd

Contents

Foreword 7
Botany in Class Five 9

The Families of Plants

1. The Plant Between Sun and Earth 15
2. The Dandelion 18
3. Fungi 21
4. Algae 25
5. Lichens 28
6. Moss 31
7. Ferns 33
8. Conifers 36
9. Trees and the Earth 39
10. Flowering Plants 41
11. Lower and Higher Flowering Plants 44
12. The blossom 47
13. Pollen 50
14. Flowers and Butterflies 53
15. Caterpillars and Butterflies 55
16. The Tulip 57
17. Seeds and Cotyledons 60
18. The Rose 63
19. The Rose Family 66
20. The Cabbage 69
21. The Nettle 72

The Plants We Use
22. The Oak	77
23. The Birch	81
24. The Palm Tree	84
25. Tea, Sugar and Coffee	87
26. Grass and Cereals	91
27. Leaves and Blossoms	95
28. Bees	99
29. Life in a beehive	102
30. The Spirit of the Bee	105
Index	109

Foreword

Charles Kovacs was a teacher at the Rudolf Steiner School in Edinburgh for many years. The Waldorf/Steiner schools sprang from the pedagogical ideas and insights of the Austrian philosopher Rudolf Steiner (1861–1925). The curriculum aims to awaken much more than merely the intellectual development – it seeks to educate the whole being of the growing child, that each may develop their full human and spiritual potential.

During his time as a teacher Charles Kovacs wrote extensive notes of his lessons day by day. Since then these texts have been used and appreciated by teachers in Edinburgh and other Steiner Waldorf Schools for many years. This book represents the way one teacher taught a particular group of children, other teachers will find their own way of presenting the material.

There is an introduction by Charles Kovacs based on what he told parents about teaching botany in Class Five.

Astrid Maclean, Edinburgh 2005

Botany in Class Five

If we consider the stage of development at which the children of ten or eleven are, an analogy from history may help.

It would be a mistake to assume that human beings have always thought logically, intellectually, as we do now. In fact, there is a definite point in time when logical thought as we know it begins – when scientific inquiry as we know it begins – when rational thought, speculation, analysis, makes its entry. The time when the faculty of rational, logical thought makes its appearance is ancient Greece, the time of the Greek philosophers, Socrates, Plato, Aristotle.

Older civilizations – India, Babylon, Egypt, even Greece itself before the age of philosophy – cultivated a different faculty. In India, Babylon or Egypt a *myth* was as valid an explanation of the world, of natural phenomena, as a scientific explanation is valid for us. Now a myth, a story of gods, heroes, monsters is not just the wild roaming of fancy. A true myth of antiquity is something which does contain poetic fantasy, but also a logic of its own. One could say that in the ancient civilizations, these two faculties, fantasy and logic, are not yet separated, they are still a unity. The Greek philosophers, the first rational, logical thinkers, appear at a certain point in time, simply because that is the time when fantasy and logic became separate and independent functions of the human mind. That is also the time when poetry emerges as a separate art.

Now what took place in human history on a large scale has its counterpart in the development of the individual human being. Around the age of eleven or twelve there is a beginning of that great divide where fantasy and logic part company.

It shows itself, for instance, as curiosity. Some children

become fascinated by facts, any facts, of science, geography, history. It shows itself in other ways: some children get argumentative. They like to argue for argument's sake, but it is in reality no more than a new toy they have discovered; they like to play with this new toy.

Another side of this development is an enhanced feeling of one's own personality. A child at this stage is inclined to say: "I like this" or "I don't like that" with quite a new emphasis – it is not at all like the "likes" and "dislikes" of younger children.

Even the naughtiness of children at this stage is different. Younger children can be naughty because they can't help it; they follow some urge they are unable to control. But children of this age are much more deliberately naughty. It is much more like a scientific experiment: "How much can I get away with?" "How far can I go?" And all this is connected with the process of *awakening*. For the emergence of logic, this parting of the ways between logic and fantasy, is an awakening. In the deliberate naughtiness, in the argument, in the enhanced feeling of one's own importance, in the curiosity, in all this the child experiences an awakening, as from a dream.

But there is one thing we must keep in mind. At this stage there is only the beginning of the process, it is not yet completed. If at this stage you give children *just* facts, the bare facts of science or of geography, then you are not really helping the growth and development that takes place in the child. They need the facts, they want the facts, but they must be linked in a way that still satisfies the feeling – the fantasy – the poetry in the child. Give them only the bare facts, then their fantasy, imagination and original creative ability, dies or withers.

That is why the botany lesson was given in the form set out here. The plants are given in the sequence of the evolutionary system, beginning with lower plant forms, fungi, algae, which have no flowers, pollen, seeds, and leading up to the flowering plants.

But the idea of evolution, of higher or lower plants, would be meaningless to the child. It has to be brought near to them by comparing the plant families with their own development. By comparing fungi with the baby stage, algae with the toddler making its first steps, monocotyledons with the first years at school – by these analogies the children are introduced to evolution (although I never mentioned the word) in a way which makes them aware of their own evolution, their own development, something they can *feel*, not merely know. Of course such analogies contain an element of fantasy, they are poetic analogies, but this is exactly what the child still needs.

I must also say something about the place of the botany lessons in the science curriculum of the school. We proceed in science as we proceed in geography – that is we begin with what is the nearest and then move farther and farther out. In geography we start with the local city, then the local country, the neighbouring countries, the continent and then the other continents. In science we begin with the kingdom which is nearest to human being – the animal kingdom – at age nine and ten (Class Four). Next we take the plant kingdom at age ten and eleven (Class Five). The mineral world, that is geology and the first steps in physics, come at age eleven and twelve (Class Six). Physics and chemistry come at age twelve to fourteen (Classes Seven and Eight). Mechanics, which is farthest away from our feeling, comes at age fourteen (Class Eight).

Charles Kovacs

The Families of Plants

1. The Plant Between Sun and Earth

In the winter when we have a particularly long spell of cold days, it is not only human beings who suffer; the plant and animal worlds are affected too. The birds are later than usual with their nest building, and the flowers and trees, in gardens and fields, on the hills and by the streams, are all kept waiting for the warm light of the sun. Just think of all the countless seeds under the earth, waiting for the light and warmth of the sun.

Imagine all these millions and millions of seeds held inside the earth during the cold season when there is snow, ice and cold winds, but they are safely kept inside. And now imagine for a moment that each seed is a tiny light, and if you can imagine that we could see through the earth, it would look as though there were millions of stars. During the winter, the earth would look like the starry sky.

If you struggle to understand something and suddenly you grasp it, you feel "I've got it!" it is like a sudden spark of light. This is thinking, real thinking, when you feel such a spark of light within yourself.

If you are really attentive then right through the day there are these sparks of light in you, but that is only if you are really awake. When you are really attentive, really awake, you are like the earth in wintertime with its millions of sparks of light, the stars. We are more awake in winter, and more asleep in the warmth of the long summer holidays.

But what about summer? In summer all the flowers come out of the earth, they grow towards the light and warmth of the sun. There are no longer any "stars" inside the earth. All the blossoms, the flowers in their lovely colours, are outside in the light and air.

We can compare this with sleep. When you fall asleep you can really feel that your thoughts are slipping away, and it is a good thing that they do, otherwise, if your thoughts stayed with you, you could not sleep at all.

Sometimes when people have worries: something they are thinking may be bothering them, and so they cannot sleep. You see, then, how important it is that our thoughts disappear when we want to sleep.

These thoughts are not really gone, however, for the next day, something you have truly understood, is back again. During the night it was not inside you, otherwise you could not have slept; somehow it was *outside* you, just as the flowers come outside the earth in the summer.

It would be a wonderful sight if we could see other peoples' thoughts when they sleep, as we see the flowers! Some people with wise and beautiful thoughts, would be surrounded by roses and lilies, and thoughts which are not so clever might look like mushrooms; kind thoughts would be like flowers with a sweet scent, like violets; and unkind thoughts would look like stinging nettles.

Even though we can't see their thoughts, you only have to look at a sleeping person and you can see that what was in him or her when awake, is now outside. When we are asleep – we are like the earth in summer. When we are awake – we are like the earth in winter.

In spring, as we know, all the plants begin to grow. However, plants grow not only above the earth, where we can see them, they also grow below. One part – the stem, the green leaves, the blossom – grows towards the light. This part of the plant loves the light, and needs the light, and without light it would not grow, it would die.

But the other part of the plant, the root, grows deeper and deeper into the darkness, into the earth; it loves the darkness

1. THE PLANT BETWEEN SUN AND EARTH

and it needs the darkness. If you expose the root of a plant to light – if you make a hole in the ground so that light falls on the root, for example – the root would die and so would the whole plant.

The plant needs both the light and warmth of the sun and the darkness of the earth. In the winter the earth is stronger, and in the summer the sun is stronger, but even in summer, the roots seek the darkness of earth; the plant is a child of sun and earth. Just as human children have fathers and mothers, so the plants have the earth as mother and the sun as father. They live their lives between sun and earth.

2. The Dandelion

If you walk over a field in spring or summer you are quite likely to see a common wild flower. And when you see it, you can hardly help it, you want to pick it up and blow at it; and when you blow, little white stars come off and fly away in all directions. Of course, this plant is the dandelion.

If you had walked across this field a few weeks earlier, however, then the dandelion would have looked quite different; it would have had no little stars that fly away, but many golden-yellow petals instead. This golden blossom of the dandelion looks like a little sun. And this golden blossom of the dandelion also loves the sun; it only opens up when the sun rises, and when the sun sets, or even when there is a dull day and the sky is overcast with clouds, the dandelion flower is closed. The dandelion flower is only open when there is sun, and at night or on a cloudy day it is closed.

Later on, there are the little stars that fly off, but earlier there is the golden-yellow flower that greets the sun, and earlier still, what would you see in the field – just green dandelion leaves. They have a curious shape these green leaves, with sharply pointed fringes. People thought these sharp points looked like lions' teeth, which in French translates to *dents de lions*. This became "dandelion."

So you can see three stages of the same flower, the dandelion. First it is just green leaves, then it is a golden-yellow blossom, and then it is fruit, because these flying white stars are the fruits of the dandelion.

Now we shall see how the dandelion changes from green leaves, to flower, then to fruit.

Early in the year, when the green, lion-teeth leaves arrive,

2. THE DANDELION

it is not very warm, as the warmth and light of the sun are not very strong yet. But something else helps the plant when it comes out of the ground. It is always around you, but you only feel it when there is a wind – it is the air. It is the air, together with the light of the sun which makes the green leaves.

But it gradually gets warmer, and the heat gets stronger, and the warmth and light work on the leaves, and through the warmth of the sun the top leaves are transformed, they are no longer green, but become the golden-yellow petals of the blossom. So it is no wonder that blossoms like the dandelion follow the sun; that they open up when the sun rises, and close when it gets dark. It is not so surprising, because the dandelion's coloured petals are made by the warmth of the sun.

However, the sun does not just give its warmth to the flower-petals, it also gives it to the earth from which the plant grows. Imagine it is a warm day and the sun shines for many hours upon a stone. Of course, the stone gets warm, and if you hold your hand to it you feel the warmth of the sun coming back from the stone.

But the whole earth from which the plants grow, does the same as the single stone, only it takes much longer. The warmth of the sun goes quite deep into the earth – several feet deep – and then the warmth comes back from the earth and into the air, just as you can feel the warmth coming back from the stone. But it takes much longer coming back from the earth, days and weeks, and it is not such a strong warmth that you would easily notice it.

For the plants, though, the warmth that comes back from the earth is something very real and very important. For the warmth coming back from the earth, rising from the earth, forms the fruit – in the dandelion's case, the little "stars" which fly away when you blow on them.

So while the sun is still quite weak the air and light forms the green leaves, then the heat of the sun gets stronger and forms the coloured petals, after which the heat of the sun lessens, but warmth, heat, comes back from the earth and forms the fruit, the little flying stars.

That warmth coming back from the earth lasts quite some time, even through the winter. Human beings would hardly notice it, but during the winter the green dandelion leaves don't stand up, they lie flat on the ground, and in this manner get some warmth from the earth. For the dandelion leaves this warmth coming back from the earth is very real and important.

So we have:

green leaves – air and light

blossom – warmth of the sun

fruit – warmth coming back or reflected by earth.

And what about the root in the earth? The roots are helped by water. You know that when you water a plant, it is no use spraying the leaves or blossoms, as they can't take in water. Plants can only take in the water they need through the roots – that's why you soak the earth around the plant with water, so that it goes to the roots and from the roots to the leaves, and blossom.

And so, air and warmth, and earth and water work together and give the plant leaves, blossom, fruit, and down below, roots.

3. Fungi

Each of the four elements play a part in how the plant grows: water and roots; air and leaves; sun-heat and blossom; heat coming back from the earth and fruit. But not all plants have leaves, blossoms, fruit, or even roots; there are many plants that are without proper roots, proper leaves, blossoms, or fruit.

You recall that when you really understand something, it is like an inner light. And as you grow up and learn more and more, there will be more and more of this inner light in you, and you will know and understand more and more. Now these plants have no inner light, their light comes from outside, from the sun. But the sunlight, too, is full of wisdom, and in the same way that human beings learn from their inner light, so the beings of nature, the plants, learn from the sunlight.

Plants like the dandelion or the rose – plants that have roots and stems and leaves and blossom and fruit – they have learned all the plant wisdom they can learn. Other plants without leaves, blossoms or fruit have not learnt enough. Which plants do not learn enough from the sunlight?

When you are ten or eleven years old, you can read and write and do arithmetic. When you were in the kindergarten you could not do any of these things. And before that when you were a tiny tot, you were not even clever enough to be in the kindergarten, and before that, you were a toddler who could barely walk. And even before that you were babies, you could not speak yet, you could only make funny noises and you could not walk, you could only wiggle your tiny arms and legs.

There are plants which are like babies, others are like toddlers, others are like young children, but plants like the dandelion or the rose, plants with beautiful blossoms, with leaves and fruit, they are almost like adults.

You can't remember what it was like to be a baby, because a baby is asleep most of the time – no one can remember what it is like when they are asleep. And the baby – even when it wails or wiggles or drinks, does not know what it is doing – even then, it is more asleep than awake.

The plants that are like babies, that are more asleep than awake, are mushrooms, or as they are called in botany, the fungi.

What does a baby enjoy? You can see it enjoys drinking its milk, nothing else. It would not want to be outside and play. Mushrooms do not like to be out in the open, in the sunlight, they prefer the shade where it is dark; they grow where it is damp and shady and keep away from the sunlight. Green leaves and blossoms only grow when there is sunlight, but the mushroom, the fungus, has no leaves and no blossoms, so it is the baby of the plant world.

Think of a blossom, of a flower that opens towards the light, towards the sun. If you look at the "cap" of a mushroom, it does the opposite: it turns against the light and opens towards the earth, towards darkness. The real flower turns towards the light, whereas the mushroom's cap turns away from it.

On the underside of that mushroom's cap, between the "folds," there is a fine dust. It is not seed because seeds are made by the sun, and it is not pollen, the fine dust in flowers. The tiny grains of dust that fall from the underside the mushroom are called "spores."

When such a tiny spore falls to the ground, a new mushroom does not grow right away, something quite different happens. From the spore a very fine and white thread grows, and from this one thread other threads grow, and from these threads others grow; if we could see it, it would be like a tree of fine threads – but a tree that grows and spreads under the earth in darkness.

3. FUNGI

This tree of fine, white threads, therefore, is the mushroom plant. And from this plant, or small, underground "tree," a kind of "fruit" grows, or as it is called, a "fruiting body," and that is the mushroom that you can see growing above the ground. The mushroom is a strange kind of "fruit" that comes from a tangle of white threads below the ground.

When you pick up a mushroom and you see the tiny threads at its "foot," these threads are not roots, they are bits of that "tree" or "plant" below. The mushroom has no roots – it is such a baby amongst plants that is has no proper roots, it has no green leaves, it has no blossom, and the "cap" does the opposite of a blossom, it turns away from the light.

The "proper" or higher plants, the plants that grow in the sunlight, have both blossoms and fruit. The mushroom is such a baby that it can't distinguish between blossom and fruit, so that what you see is actually the blossom and the fruit at the same time. But it is a blossom that turns away from the light and a fruit that shoots up from the darkness, a fruit not ripened by the light and warmth of the sun.

There are even "mushrooms" that grow below the ground, and these are called truffles. People like to eat truffles, but because they are under the earth truffle dogs (or very occasionally pigs) are used to sniff them out. The dog starts digging when it smell truffles under the ground, and then the people take the truffles for themselves – the dog must be satisfied with ordinary dog-food.

You can see how the fungi or mushrooms cling to Mother Earth, just like tiny babies, and you know that babies grow very quickly – in the first months of life you grow much faster than you are growing now. That is also something the plant-babies, the fungi, do. They grow very fast: one day you see nothing, then after a good rainfall mushrooms appear, several inches high. You will never see a flower grow so quickly.

A flower learns to produce lovely coloured petals and a

sweet scent from the sunlight. The mushrooms have neither petal nor sweet scent – they do not learn from the sunlight, and stay babies forever.

4. Algae

As you know mushrooms or fungi are the babies of the plant world. They never get any further than the baby stage, compared with the other plants that come out into the light. Remember, the "real" plant stays away in the darkness of the earth, only the fruiting body (or mixture of blossom and fruit) comes above ground, and even this prefers shade.

We come to another kind of plant now, one that we can't compare to a tiny baby that is more asleep than awake. We must think of the baby when it begins to stammer its first words and stand upright – standing upright, as you know, is quite an art.

You don't remember that time in your own life, the time when you began to say words, the time when you tried to get up on your legs. You can't remember it because you were still not awake enough at that time, more awake than before, but still not very awake. However, you have seen other children, brothers or sisters, at that stage.

Plants don't walk and don't talk, but there are plants that have got a bit more plant-wisdom than the fungi, plants that have gone a stage further than fungi. Now what do you think the first step in plant-wisdom is, the first step which the fungi can't take? What is the first thing you notice that mushrooms do not have? They do not have leaves, green leaves.

The next plant we come to does have leaves – they are not always green leaves, they can be brown, yellow, red – but they are leaves nonetheless. If you have ever been to the seashore after it has been a bit stormy, you will certainly have seen long branches and leaves of seaweed. And you remember that the leaves and stalks of seaweed feel quite differently to the way land-plants feel – there is no strength in them.

The proper, botanical name for seaweed is algae. The stalks and leaves of the algae cannot stand up, they cannot rise up from the earth as the dandelion leaves do; and although they cannot do it when they are on dry land, the seaweed, the algae, does stand up when held up on all sides by water.

A baby begins to stand up by holding onto something or onto a parent's hand. The algae, the seaweed, grow under water and are held up by the water, but on dry land they flop. So you see they are at the stage of little children who still need help to stand up, and they remain on this stage all their lives just as mushrooms remain babies all their lives.

How much "plant wisdom" have the algae learnt though? Just enough to have leaves; the sunlight has taught them to make leaves. They have no flowers, they do not even have real stems – what looks like a stem is just a narrow part of the leaf. A real stem stands upright like the sunray that makes it, but algae leaves and stems just float in the water.

Algae have no stems or flowers and they have no proper roots either. A real root is strong and hard, and goes deep down into the earth. Where the seaweed holds onto the seabed, however, it is only a part of its leaves that grips the rocks. It is no wonder, therefore, that during a storm a lot of algae are torn from the seabed and tossed ashore, because they have no real roots to hold them fast.

There are many kinds of seaweed, and if you dive down to where they grow, or look down when the water is clear, it looks as though there are fairy-forests, fairy-gardens, and fairy-meadows of strange plants down there. And some of these fairy-plants look like trees, like strange flowers, like fruits, but they are not trees, flowers, or fruit at all. As I told you the algae only have leaves, but some of these leaves copy or imitate the real trees, flowers, and fruits. This is just like the little toddlers

4. ALGAE

who like to imitate and copy what older people, older sisters, brothers or parents do.

In the same way the algae imitate the higher plants that grow on dry land, they haven't got the strength to produce real flowers and fruit, but they copy them. The algae, the seaweeds are not quite so asleep as the fungi; they love the light, but they can only stand upright in the water, and if you remember the spongy feel of seaweed, you know the difference between the algae and the other plant-children.

5. Lichens

We have looked at two plant families, the fungi and the algae. We compared the fungi with very young babies who are more asleep than awake, and the algae with toddlers who start to stand up and talk. These children of the plant-world have not learned to produce certain things that other plants have:

Fungi have no leaves, blossoms, fruits, or roots

Algae have no blossoms, fruits, or roots, but they do have leaves; these plants consist of leaves.

When children get older, they can stand upright, but they can only make little steps; they can speak, but only short words. You may not remember it, but there was a time when you could not have said a long word, like "geography," or "denominator." It was a time of little steps, and when you had made a few little steps you sat down, and were very proud of what you had done. It was also a time of little short words, and that was quite enough for your needs.

We have two plant families that are just like children when they make tiny steps and use short little words. And both kinds of plants are very small indeed. The first of these tiny plants you can often see on old stones, old rocks, and also on the bark of old trees. It looks as if the rock or the tree has been sprayed with grey or green paint, but if you look closer you can see that it is covered with little greenish-greyish scales which are actually tiny leaves. These plants are called lichens. There is also a kind of lichen that hangs down from branches and looks like a beard that is made up of tiny stalks and branches.

They are strange children, these lichens, as they cannot grow in good soil, they can only grow in the crack of a hard surface – on a rock, on the bark of a tree; they have no proper

5. LICHENS

roots, they can't go deep down into the earth. The sunlight has taught the lichens to make tiny leaves, but they don't live in the water like algae. However, these leaves cannot stand up generally; the ones that can stand, can only do so to a very small degree. Some lichens are more like stalks, and some are more like leaves, but they can't produce flowers or fruits.

And what about roots? You recall that the "higher plants," the more "grown up" plants get the water they need through their roots. Like the algae, lichens do not have proper roots, just tiny threads that enable them to hang on to the rock, or the bark, or to the wall.

How do lichens get water? These strange, tiny plants do what the higher plants cannot do. They absorb water through their leaves. When it rains, even when the air is damp with fog or mist, the lichen-leaves absorb water.

Lichens are very hardy plants. They can live for months and months without water; they can get as dry as dust – it does not matter to them, they can wait – and with the first drops of rain, they spring into life and begin growing again.

In the far north there are kinds of lichens that are frozen for many months, but when the ice melts, they grow and spread happily. One of these hardy lichens is called "reindeer lichen," and it covers the hard frozen ground where it grows like a grey cushion. As the name tells you, reindeer eat these lichens during the long cold winters of the north.

Something interesting that scientists have discovered about lichens, is that they are made of not one kind of plant, but two plants, working closely together as though they were one. One of the plants in this couple is more a child of the sun, it is really a kind of alga, and it forms the green part inside the little leaves. Its companion is more a child of the earth, a tiny fungus. The fungus holds onto the stones or bark and grows around the alga, keeping it safe, enclosing and protecting it. The alga gets nourishment from the sunlight and shares this with its friend the fungus. And so, by forming a partnership, both of them can live in places where neither could live alone.

How do lichens get about? When we see lichens growing on stones, where do they come from? Lichens produce a grey green dust of tiny leaves that have simply crumbled into dust. The wind and the rain carry the dust about to other stones or bark where it falls into tiny cracks and begins to grow. Lichens have no true seeds.

The lichen is a tiny, but sturdy plant.

6. Moss

The fungi were like very young babies, the algae like toddlers who are just learning to stand up, and the lichens like little children who begin to take their first steps. The next plant we come to is also like a little child who takes its first little steps, who says its first words. It is also a tiny plant, so tiny that many of you hardly pay any attention to it, yet you cannot walk in any wood or forest without meeting it, without feeling it under your feet. It is moss.

Like the lichen, the moss is a tiny plant, and while a tiny lichen leaf doesn't have much shape – it is as though somebody drew it rather carelessly – each of the tiny leaves of moss, appear as though they were created with utmost care.

If you look at moss closely you will see that it consists of many separate plants. Some look like little fir trees, others have little round leaves. There are a great number of different kinds of moss.

When you are in a wood, you look up to the tall trees around you and see many different kinds of trees. But underfoot in the moss, there are tiny forests, a whole world of plants as wonderful as the tall world of the trees.

One or two trees don't make a wood; there must be many to make it a wood. Every patch of moss is like a tiny wood. The tiny plants of moss always grow in company, they are always in a little crowd. That is why the moss lies like a cushion over the earth in a wood, a soft cushion made up of thousands of little plants.

The tall trees of the wood need this cushion of little trees down on the ground. These soft, green cushions of moss do something that the lichen also does: they absorb water like a sponge, and hold on to it. If there were no moss, then when it rained, the rain water would flow away, and soon after that the ground would be dry. The roots of the tall trees would not find enough water. However, these soft spongy cushions of moss keep the ground damp enough for the tall trees to get the water they need. So these little moss-plants are not selfish, but generously help the whole forest.

We know that moss is green, so we already know that moss – like algae and lichen – has learnt the first lesson of plant-wisdom, that is, to make green leaves. But the moss-plants keep in the shade, and they also keep close to Mother Earth. They lift themselves up a tiny bit, they have upright stalks. But they don't stand tall like tulips or dandelions, for they are still like young children within the plant world. And as they like the shade, the sun cannot teach them much, and their leaves remain small. But these tiny leaves stand up, without help.

Of course, the moss-plants have no real blossoms, no real fruit, but they posess a kind of imitation of fruits and blossoms. Some of the tiny leaves on top of the stem turn yellow, so it looks like a small flower, but real flowers don't consist of green leaves that turn yellow, they are special leaves, called petals. And if you have ever seen the little round capsules where poppies keep their seeds, the moss-plants imitate that too, but it is not a real fruit as the poppy's capsule is.

The moss leaves take in water, so they do not need roots. They have no real roots, only very fine, short rootlets. It is easy to lift a patch of moss from the ground. It comes away readily because the only things attaching it to the ground are fine threads. The mosses, like the lichen, are really the toddlers of the plant world, tiny plants, that just have the bare beginning of roots, and imitations of blossoms and roots.

7. Ferns

We have heard of the babies and the toddlers of the plant world. Before we come to the next kind of plant, we shall once again look at the next stage in a child's life as it grows up.

When a little child first begins to speak, and hears herself called, for example, Mary, then at first she will say "Mary wants this" not "I want this." Later comes the great discovery that when you speak of yourself, you don't call yourself by your first name but instead you say "I." That is a very great step in awakening and usually one of the earliest things you can remember. This first awakening, then, happens around the time you say "I" instead of calling yourself by your name.

It is a great step in waking up, and as we grow up, more and more awakens within us. Even now, only a small part of you has woken up, and more will awaken as you continue to grow up.

And there are plants that are like children when they call themselves "I" for the first time. These plants are mostly to be found in the forest, and they too only have green leaves, but they are quite tall and are very beautiful.

They are so beautiful that florists, when they make up bunches of flowers, often add some of these green leaves to a bunch, to make it nicer. And you must work with real care and love in order to draw these beautiful leaves. These plants are the ferns. This beautiful green leaf is the same as saying "I." Just think how clumsy dandelion leaves are compared with ferns!

If you watch ferns growing you would hardly expect them to become such wonderful leaves. What you see at first looks like a little snail's shell, and it is not even green, but brownish. As these little snail-like things grow, however, they unfurl and it is quite a wonderful sight, to see these rolled up little things unfold and become lovely, tall, fern leaves. One must really study a fern leaf to see how it is made: a strong middle rib, side ribs which carry little leaves. These get smaller and smaller towards the tips of the side ribs.

The higher plants – for example, roses, carnations, all the other plants which have flowers – give all their attention to the blossom and don't pay too much attention to the green leaves. But ferns are plants that have no blossom, and all the strength and wisdom that the sunlight gives them, is used to make these beautiful green fern leaves.

And they do something else: these fern leaves in one way imitate true blossoms and flowers. Just as the petals of flowers stand in a circle, so the fern leaves grow in a circle. But, of course, they are just green leaves growing out of the ground, not flowers or blossoms.

These beautiful strong ferns are already quite different from the tiny lichen and moss, from the soft spongy seaweed, and from the fungi. The ferns are true, beautiful, green leaves, they are much more awake than the others. Just as awake as children are when they say "I" for the first time.

However, they still cannot produce blossoms or fruits. So how do new ferns grow every year? If you look on the back of each leaf, you can see brown dots, and if you put a fern leaf on blotting paper (in summer), the next day you will find that the brown dots have left a print on the paper. This brown dust is like the spores of the fungi. But there is a difference: from the fern-dust little green scales grow, and later, from these green scales, new fern-leaves eventually grow.

Ferns have a relative, a plant that is, one could say, a cousin of the ferns. But that plant does not bear much likeness to the fern. It has no beautiful leaves, in fact it has hardly any leaves at all, but beautiful long stalks, and the whole plant looks rather like a very small fir tree. It is called the horsetail.

7. FERNS

The long stalks of the horsetail are hard but brittle, as though they were made of very fine glass – if you shake a horsetail plant you can even hear a crackling. But these strange stiff proud fellows, the horsetails, are relatives of the ferns with their beautiful curves. The ferns are leaves without stems (they grow from the ground, from the roots). The horsetails are stems without leaves.

Neither ferns nor horsetails have blossoms. The ferns put all their strength into leaves, the horsetails put all their strength into stalks, and neither can produce blossoms.

8. Conifers

We compared the plants and growing children: from fungi, algae, lichen to moss was like babies, toddlers and little children who say "I" for the first time.

We are about two or three years old when we learn to say "I," and in nature that is the stage the beautiful fern-leaf is at. After that, is the stage before we go to school, the nursery stage, when we are three, four, and five years old. In the years before you go to school, you grow more and more awake. A child of four is no longer a baby, is no longer a toddler. They enjoy a simple story, and like to hear it over and over again. Four-year-olds are often clever enough to count – many can count up to 20 – but they cannot add, subtract or multiply.

We saw that plants learn from the sun. And the most wonderful thing a plant learns from the sun is to make a flower, a blossom. But only plants that really learn from the sun, only the plants that love to learn from the sunlight, can make true blossoms, true flowers.

There are plants that are like a four-year-old child, they are like the child before it comes to school. They are more awake than babies, than toddlers, than children who can say "I," but they don't learn as one learns in school, and so these plants still cannot make any flowers, they do not have blossoms.

We can consider the first awakening to be green leaves. The fungi don't have these, but the algae, lichen and moss have, and the ferns have the most beautiful green leaves, but they are still just green leaves. The horsetail has the most beautiful green stalks, they have longer sections at the bottom and they get smaller and smaller towards the top. That might remind you of another plant. Look at how like a little tree the horsetail

8. CONIFERS

is, and how regularly the side-branches grow from the trunk of that tree. An oak-tree or an apple tree's branches do not grow like that.

But fir trees do grow like this. Fir trees have a straight trunk that gets thinner and thinner towards the top, and the branches grow from the sides as regularly as the branches of the horsetail. You could say they grow on separate floors – first floor, second floor and so on – just like the horsetail branches.

You could say the fir tree is a kind of horsetail that has grown much taller, and to grow taller it had to develop into wood, as a green stalk could not grow that high. The leaves we call needles, as they don't spread out like flat leaves; they are more like stalks, as the horsetail leaves are. Pine trees and fir trees as well as the larch belong to this family of plants.

These trees are all much taller than the horsetail, but they have not learned much more than the horsetail, for all these trees with needles for leaves have no blossoms, and that's why we say they are like four-year-old children, before they go to school.

These trees have cones, however, and we shall learn more about them. When other trees have blossoms, the fir trees and pine trees grow many, many tiny trees on their branches, each with a tiny, straight trunk in the middle. Around this tiny trunk the scales grow, and the whole cone develops. If you take a closed fir-cone (or pine-cone) home and keep it for a couple of days in a warm room, the scales open up like little doors and you will see that from under each scale two little winged seeds protrude, ready to be blown away by the wind.

In spring when the pine-cones are still small, they are sometimes red and stand like small red candles on the branches – they are like blossoms. Later when the seeds fly off them, they are like fruit. The needle-bearing trees have not learned to distinguish between blossom and fruit yet; they do not have separate blossoms and fruit, only cones that are both. These trees are called conifers (cone-bearers). The proper name of the needle-bearing tree is conifer.

In early spring the horsetail also grows a tiny cone on a single stalk, but it is not a proper conifer, only trees can be called conifers. The conifers do not have "proper" blossom and fruit, but they don't send out a fine dust either, like ferns or mosses of fungi, they have proper seeds.

Conifers also have a sticky substance called resin inside them, which flows out if the bark is cut, and it has a lovely smell, especially when you burn it. The conifers do not grow proper blossoms, but all the scent that would go into blossoms if they had them, goes into the resin instead. You could say that there is an enchanted flower in the resin, and when you burn a fir-branch, the enchanted flower that is in the resin is released in the flame and scent.

9. Trees and the Earth

We have learned about children who are four years old, who still don't go to school, and we have compared them with conifer trees, the straight fir trees and the pine trees. If you compare a tall fir tree with a dandelion, it is quite small in comparison. Yet the dandelion knows more; it has learned more from the sunlight than the fir tree – the dandelion has proper blossoms, and later on it has proper fruit, proper seeds.

You may think it is strange that the tall fir tree is like a four-year-old child before going to school, and the dandelion, which is so tiny in comparison, is like a much older child – it is much more awake than the tall fir tree. The size of the tree compared to the size of the little flower does not matter, however. To understand it, let us think of another tree, for instance, a cherry tree or an apple tree in full blossom – it is a lovely sight! The cherry tree or the apple tree in blossom, in itself, is like a little flower garden, except that all the flowers are the same kind of flowers. Each tree is a flower garden of the same flowers at blossom time.

And in winter these same trees are just as bare as the earth under them, the wood, the trunk and branches; in winter it is just like the earth itself! When spring comes there are green shoots coming from the earth, and green shoots coming from the wood – the wood has become like the earth again. The wood behaves like the earth.

But there is a difference: the plants down on the ground each have their own root in the earth; the leaves and blossoms on the tree have all the roots of the tree on which they grow.

The plants that grow on earth send their roots into the earth, and the earth makes the roots hard, and some roots even become wood. It is the strength of the earth that hardens the roots so that some of them become as hard as wood. It is not the sun; the sun makes the fine flower.

You can now understand that where a tree grows, where you see a hard wooden trunk rising from the ground, it is the earth-force, the earth-strength itself which goes upwards; it is as though a hill were pushed upwards, pushed up until it becomes a tree.

Because the wood of a tree is really the earth pushed upwards, when a tree gets old and begins to decay, to crumble, what happens? If you put your hand into a hole in an old decaying tree, you find wood that has crumbled into soft earth. So it is no wonder that the wood behaves like the earth – it is bare when the earth is bare, and becomes a little garden when garden flowers grow from the earth.

The fungi and the lichen and the moss all know that the wood of the trees is a kind of earth; that's why you find fungi and lichen and moss growing on trees – they are children of earth and they grow just as happily on wood as on the earth. But there are also insects, beetles, and worms that live in the earth, and they too live just as happily in the wood of trees – they are all aware that wood is a kind of earth.

Now you become aware that the green leaf, the blossom, and the fruit is the real plant in the tree, and the tree itself, the wood, is only a great hill pushed upwards. Think again of the fir tree and the dandelion. The fir's real plant is the needle and cone, not the wood. And on that "real" plant, the needles are very simple leaves compared to the dandelion leaves. And the dandelion flower is not only one blossom, but hundreds of blossoms growing together.

You might ask why there are different kinds of wood. It is because the green plants that grow on the wood need different kinds of earth. It is the same with plants that grow below the ground – many of them need a special kind of earth; and the little green plants on the trees need their own special earth, and that earth is the particular wood of that tree.

10. Flowering Plants

Trees are really a kind of earth-hill, pushed upwards, and the leaves, blossoms, and fruit, that grow on the branches, are like other plants growing down on the ground. The only difference is that the leaves and blossoms up in the tree's "garden" don't have individual roots, instead they share all of the tree's roots.

But the first trees we heard about, the conifers, do not have true blossoms, they have green leaves like tiny needles and on most conifers they remain green all year round, summer and winter. That is why these trees are also called evergreens. They do not have true blossoms; they have not learned from the sun how to make blossoms, and therefore, they are like children who are four or five years old, who have not gone to school yet.

Now we come to the stage and to the age when children go to school. Now we come to the time when you begin to learn to read and to write. The inner light of thought had already been at work within you. When you could count up to twenty or thirty, could do little drawings, and called yourself "I" for the first time, the inner light had been working within you.

But once you came to school, this inner light had to achieve much more. To learn to write and read needs much more effort by the inner light in you, than anything you did before. Just think of adding and subtracting and of the multiplication tables – it needs quite a bit of inner light to learn things like that! If you love this inner light, then you are like the plants which have beautiful flowers; you are like the red and pink and blue and white and yellow blossoms which our eyes enjoy; you are like the flowers that fill the air with lovely scents.

All the flowers and blossoms that open themselves so joyfully to the sun, are like children who are open and willing to learn. And of course, people who think more of mischief, who don't care to pay attention, are like toadstools that turn away from the light. And some toadstools are poisonous, and only grow where there is something dead.

The plants that bear flowers are called flowering plants, they are like children who come to school and really take in what they hear. But the flowering plants, the plants whose blossoms fill the fields and gardens with beautiful scents, do not have a school, and do not have teachers as human beings have. Their school and their teachers are the rays of the sun.

But it is not only the light of the sun, there is something else at work. At night-time when you look up to the sky, you see the countless stars, and their light also comes to us down on the earth. In the daytime when the sun shines, you may think there are no stars in the sky, but you are mistaken, because the stars also shine during the day. However, the light of the sun is so strong, so powerful that we cannot see the stars, but they are there, they shine just the same. In the daytime, along with the light of the sun, the light of the stars also shines down on earth. Our human eyes may not notice it, but for the plants, the starlight is just as real as the sunlight.

From the sun the plants learn to make a flower, but from the stars they learn to give their flowers starlike shapes. Some are like six-pointed stars, some are like five-pointed stars, some are like four-pointed stars, and some are like stars with many rays. Some plants even show their star in their fruit – like the "star" in an apple, or in oranges and lemons.

You could say that the stars on high are the heavenly flowers, the flowers of God. And our flowers here on earth are like a "mirror" that reflects the light of the heavenly flowers.

Our earthly flowers are of course, only tiny reflections, tiny pictures of the wonderful lights of heaven. But when you are in school and you really learn and take things in, then this tiny bit of cleverness, of wisdom, of intelligence that human beings absorb, is also a tiny reflection, a tiny "mirror," a picture of the great, infinite wisdom of God.

11. Lower and Higher Flowering Plants

If you want to know what real attention is, then you can learn it from the flowers. The dandelion and some plants open their blossoms at sunrise and close them at sunset. But there are other plants that are even more attentive, the sunflower for instance. The sunflower always turns its head, it's great blossom, toward the sun. As the sun moves across the sky, the sunflower follows its light right through the day.

But even smaller flowers do this. The violet also follows the sun – it turns its little head slowly so that it can always receive the sun's rays right into the heart of the blossom. The little violet is not only an attentive flower; it is also a flower that does not draw attention to itself – it hardly shows itself. And the violet also has a lovely scent, but you have to be close to it to notice it.

Carnations are quite different, they don't follow the sun so closely. They are very proud and conceited, having such pretty colours, and with some varieties their scent is so strong that, if you place violets or even roses beside them, you would hardly notice the others. Carnations are like children who are not especially attentive, but they want to draw attention to themselves, they want to show off.

The flowering plants are very much like school children. There are plants like the dandelions or daffodils, which have sturdy stalks that have enough strength to stand upright, but there are also climbing plants, which must always hold onto something else.

In the same way there are children who work on their

11. LOWER AND HIGHER FLOWERING PLANTS   45

Left: Leaf with parallel veins. Centre and right: leaf with reticulate veins.

own, and others who always need help. There are also plants that like to make themselves unpleasant, the stinging nettles, for example. There is a way of dealing with stinging nettles, however. If you grasp them firmly they won't sting you, they can't. And, of course, it is the same with children who get up to mischief – one has to "grasp" them firmly.

Just as there are younger and older school-children, so there are two groups of flowering plants, a higher one and a lower one. And you can tell these two groups by their leaves. The lower flowering plants have *simple* leaves, and the higher flowering plants have more *complicated* leaves.

The leaves have so-called "veins," and the simple leaves have veins that are more or less parallel. But, the more complicated leaves, have veins that divide and cross over.

In botany they are called parallel and reticulate veins. The complicated leaves may also have more complicated edges, like the dandelion, or rose but the simple leaves have straight edges.

You can also tell them by their flowers. Lower flowering plants have flowers that follow the six-pointed star, like crocus, lilies, tulips, iris, snowdrops, narcissus, or daffodils. The higher plants follow the five-pointed, four-pointed, or many-pointed star, like rose, violet, or carnation.

In addition, the flower or blossom of these plants, grows from a cup of green leaves, called the calyx. The plants with parallel veins have no calyx, no little cup – they can't yet distinguish between calyx and blossom.

The higher plants, with complicated leaves have strong roots. The lower plants have a bulb underneath, which is not a root but a thick part of the stalk, and only tiny roots come from this bulb. So leaves, blossoms and roots are all different in these two groups.

12. The blossom

The plants that are like children before they go to school – mosses, ferns, algae – are all plants without blossoms. And the plants that have flowers – tulips, violets, daffodils, roses – they are all like school-children. In the beauty of the blossoms, in the scent, it is like seeing what it means to learn, to make the inner light work in you. Just as it is not always easy to learn, to remember, to think, so it is not so easy for plants to bear blossoms. In fact, the flowering plants must prepare themselves before they can bear blossoms. A flowering plant – a rose, or a wild plant, the buttercup, for example – grows and grows for quite some time, while there are still only green leaves visible.

With the buttercup the leaves change even as the plant grows. The first leaves, that are near the ground, near the earth, are large, but later on the leaves that develop higher up, are smaller and finer. But they are still just green leaves.

Then a wonderful thing happens: the buttercup stops growing taller, it grows special little green leaves that are quite different from the green leaves that came before, and these little green leaves stand close together and stick together, so that they form a little green shell. If you take a knife and open that little shell just as it has formed, you would find nothing in it. But if you wait a little while, a few more days, then this little shell opens up to reveal a lovely yellow blossom! And the little shell is now like a cup, which holds up the petals. These little green leaves which make a shell at first, and then a cup, are together called the calyx which means cup. And the little leaves themselves are called sepals.

The sepals, that together form the calyx, are at first like hands in prayer, and then they are like hands holding a precious gift, and that is the flower.

The coloured leaves of the flower are called petals. But the gift that the calyx has received is not only these lovely, coloured petals it has also received something else. In the very middle, in the centre of the petals, standing straight up, there is something that looks like the sceptre of a king, or queen – it is called the pistil.

In olden times a king not only held this golden rod, the sceptre, in his right hand, but in his left hand he held a golden ball, an orb. When a king sat on his throne at the coronation, he held the sceptre in one hand, and the orb in the other – a rod and a ball.

If you look a bit closer at the sceptre that stands in the middle of the flower, you will see that, further down, it widens out and looks like an orb. In the plant it is one thing, the upper part is like a scepter, the lower part is like an orb. But, in botany they are not called sceptre and orb. The upper, rod-like, sceptre part is called the pistil and the little ball, the lower part, which is like an orb, is called the ovary.

This is not all that the calyx holds up, however. It holds up the petals, the pistil and ovary (which are really one), but it also holds up something else. Around the proud royal sceptre and orb – around the pistil and ovary – in a circle, like a crown, there stand fine stalks known as stamens. The stamens have little golden heads and these golden heads are made up of a fine golden dust, called pollen.

You recall that at the beginning of this book, you learned that the plants are children of sun and earth, and in every part of a flowering plant there is something that is given by the earth, and something that is given by the sun. You can easily see that the green pistil and the round ovary (itself like a little earth) are the gift of the earth. And the stamens, standing in a circle, with the golden pollen, are the gift of the sun.

12. THE BLOSSOM

What a wonderful gift the calyx hold up: the petals, the pistil and ovary, and the stamens with their golden pollen.

13. Pollen

In olden times a king on this throne carried in one hand his sceptre and in the other his orb, a golden ball. We know the sceptre, the golden rod, is a sign of power. But orb, the golden ball, means the earth. These kings of ancient times wanted to show that the earth was in their power, and that's why they carried these orbs.

Let us have a look at the little orb inside the flower, the little orb called the ovary. Does this ovary have anything to do with the earth? You remember that in winter, in the cold season, the seeds lie in the earth, waiting for spring. But where do the seeds come from, where is their beginning? Their beginning is in the ovary. The same seeds that will later on lie in the earth, they lie at first in the little earth, in the ovary. The ovary is really a little earth.

It would be no use if you tried to take the tiny seeds from the ovary and put them into the big earth. If you did that nothing would grow from them. These tiny seeds in the ovary need something else before a new plant can grow from them. They need the blessing of the sun.

You know that the pistil and ovary are the gift of the earth. But what is the gift of the sun? The stamens and the golden pollen, this golden crown, are the gift of the sun. The golden dust of the pollen is needed to bring the blessing of the sun to the seeds.

How can the pollen's golden dust get to the tiny seeds inside the ovary, in that case? The "sceptre," the pistil is hollow and has a sticky top. And for some plants, the wind carries the pollen from the stamen to the pistil, and so they make their way through the hollow pistil, to the seeds in the ovary.

13. POLLEN

On many flowers the golden pollen dust is not a dry dust, but a sticky dust – it will stick to your fingers if you touch it. And for all these flowers, the pollen is brought to the pistil by insects, like bees and butterflies.

Plants and insects help each other. Deep down in their cup, flowers have a tiny, tiny drop of honey or nectar. It is so little that we could not even taste it. But for a little bee, or a butterfly, this drop of honey is quite a large amount. And the bees for instance, don't only go to one flower, they visit many flowers, and collect some honey from each and carry it to the beehive. And every beehive has thousands of bees bringing honey, so the bees gather quite a great quantity of honey together.

So, the flowers give their honey to bees and butterflies; the flowers don't produce honey for themselves, they do it for the insects. But the insect also does something for the plant.

When the bee collects a little drop of honey, a good deal of that golden dust, the pollen, sticks to the bee, or the butterfly. And when the bee comes to the next flower of the same kind (it must be the same kind) they rub some of that pollen off themselves and onto the pistil, so the pollen can, once again, reach through the hollow pistil to the seeds in the ovary. And bees seem to know about this, for a bee will only visit one kind of flower, say only apple-blossoms, in one day, and the next day, only cherry-blossoms and so on. In this way the flowers give the insects honey and the insects help the pollen, the blessing of the sun, reach the tiny seeds in the ovary.

Without the pollen, without the blessing of the sun, nothing would ever grow from the seeds in the ovary. But once the seeds have received the blessing of the sun, something wonderful happens.

The ovary, the little earth, begins to grow. The other parts of the blossom, the petals, the stamens, they fall off, because their time is past. But the ovary, the little earth that has received the blessing of the sun remains, and grows, and grows and it becomes the fruit.

Inside the fruit are the seeds, and when these seeds are put into the earth, a new plant can grow from them. And so, if you cut open an apple, or an orange, a tomato, or a cherry, what you find inside are the seeds. But the juicy, sweet "flesh" of these fruits, at first, is the little "orb," the ovary of the flower. So you see, sun and earth, flowers and insects, all have to work together so that there are fruits and seeds in the world.

14. Flowers and Butterflies

We have discovered how flowers and insects work together, and many flowers and insects simply could not live without each other. If the bees and butterflies did not come to the flowers, the pollen would not reach the seeds in the ovary and next year there would be no more flowers of this kind.

But the bees and butterflies also need the flowers. Insects such as bees and butterflies are children of light and warmth. On a cold or wet summer's day they hide, they don't come out, but on a warm sunny day, the bees and butterflies are as happy as a fish in water. And these children of the light never take any food from the earth itself (just water occasionally); they only feed on the sweet drops of nectar which the sun has prepared for them in the flowers.

Flowers and insects belong together, they need each other. If you think of butterflies, and their colourful wings as they flutter from flower to flower, then think of the colourful blossoms, which are also children of light, like the butterflies. Of course, the flowers, the petals, are part of a plant and the butterfly is an insect, so they can't be quite the same, but they are very much like each other. (Many flowers close on a wet, dull day just as the butterflies go into hiding!) There is certainly a likeness between petals and the wings of butterflies: the body of a butterfly also looks like a little plant stem, and the long feelers of the butterfly, they look just like stamens. If you tried you could make a butterfly from the parts of a flower.

But if we only look at the finished flower, and at the finished butterfly, we cannot yet see how great the likeness is. We shall understand much better, how similar flowers and butterflies are, if we see how they grow. The plant begins as a seed. From the seed comes the green plant shoot (green stalk, green leaves). Next comes the bud, the little green shell, and it seems as if nothing more will grow. And then the bud, the calyx, opens and out comes the flower, the colourful blossom.

The butterfly begins with a little egg. The egg is like the seed of the plant, the seed and egg are "hatched" by the sun. Next comes the caterpillar and it grows as fast as the green plant, it eats the leaves of green plants – it is like the green plant.

But after a time the caterpillar stops growing and it does a very strange thing: it encloses itself in a skin, a fine, thin shell, which is called a "chrysalis," and for a little while nothing more seems to happen. Of course, the chrysalis is just like the bud, like the calyx before it opens. And then this "shell" opens and out comes, not a caterpillar, but a beautiful butterfly, just the beautiful flower comes out of the calyx!

> Blossom – Butterfly
> Bud – Chrysalis
> Green Plant – Caterpillar
> Seed – Egg

One could say the blossom is really a butterfly held fast, a butterfly that has been bound to the earth. And the butterfly is a blossom, a flower that has been set free so that it can flutter in the air.

Once you understand that plants – from the seed to the flower – are butterflies, bound to the earth, then you can also understand why they seek each other; the butterfly seeks the flower, and the flower loves the butterfly and has nectar for it. And you also understand why there is a likeness, a similarity between them. They are like two children of the light: one bound to the earth, and the other flutters and flies freely in the air and in the light.

15. Caterpillars and Butterflies

Flowers and butterflies are both children of the light, they love the light. On a dull, wet day many flowers close their blossoms, and their "sisters," the butterflies, go into hiding.

The butterflies, moths and also many other insects, love the light so much that they even give their life to it. At night, in summer, a light will attract all kinds of insects, and if you watch, you can see that they really want to throw themselves into the light, but they can't get through the electric bulb.

The caterpillar, hatched from the egg by the sun, also loves the light; it loves the sunlight. And if the caterpillar could fly up to the sun, it would do it, just as the insects fly to the little light in your room.

But the caterpillar cannot fly up to the sun, it has no wings, and even if it had wings, the sun is too far away. So the caterpillar does something else: from its own body it makes a thin thread, as fine as a silk thread, and it spins and weaves this thread around the sunrays. And that's how the little shroud around the chrysalis, the cocoon, is made. The cocoon is made of a fine thread woven by the caterpillar around sunrays.

And when the cocoon is finished, the caterpillar changes completely. The caterpillar has died, it has gone, and all that is left is a shell – the chrysalis. But now something wonderful happens, by the power of the sun, inside the chrysalis the butterfly is born. The caterpillar dies, but is reborn as a butterfly.

A caterpillar is not a very beautiful creature, it looks like a hairy worm, but this ugly hairy worm gives up its life; it gives up its whole body, it dies and becomes a dead shell. And from this dead shell, the butterfly, that beautiful winged child of light rises. The hairy worm dies and is reborn as the glorious butterfly.

God made such a strange creature as the caterpillar for a purpose. You see, God in his infinite wisdom speaks to us through all things in nature. God wants to tell us something by showing us the caterpillar that dies and is reborn as a butterfly.

All human beings must die. For all of us comes the time when we must die as the caterpillar dies and becomes an empty, dead shell. And just as the glorious butterfly is reborn from the dead chrysalis, so we all are reborn as spirits in the heavenly kingdom of God. Of course, the butterfly you see fluttering about is not a heavenly spirit as we shall be; the butterfly is only a picture through which God speaks to us.

When God made the butterfly he thought: "The human beings on earth are sad when somebody they love dies, but I will show them the caterpillar that dies, but it is reborn as a butterfly and this should tell them: Be of good cheer, death is not the end. Just as the glorious butterfly came from the death of the caterpillar, so you will be awakened to new life as beautiful spirits when you die."

Plants and butterflies, the things we see around us in nature tell us something about ourselves. God made them so that we learn something about ourselves. But caterpillars and butterflies were made, so that we know that when we die, we shall be reborn in the spirit.

16. The Tulip

We saw that there are two kinds of flowering plants: one kind in which the veins are in parallel stripes; and one kind in which the veins are in a network, where they look like little trees. Many, though not all, of the plants whose leaves have parallel stripes show their flowers quite early in the year. Think of the snowdrops which come so early in the year, then the daffodils, hyacinths, narcissus – they all come early in the year, before the sun has reached full power. (Roses are quite different, however, as they will only come in June when the sun is in full power.)

All the parallel-veined plants are ruled by the six-point star. Another plant that belongs to this group is the tulip. The tulip is also rather early, and you could even make tulips grow long before springtime in a warm room. The reason why tulips can grow so early is because they don't have to grow from a seed every time, they grow from a bulb instead.

A bulb is a curious thing: outside there is a leather-like brown skin, and inside are white skins, one inside the other. Below there is a flat disc, and from this disc, real fleshy roots grow – not hard ones – and they do not branch out, for the bulb itself is not a root. The skins that form the bulb are nothing else but leaves; they are leaves that have been kept underground, and because the sun cannot reach them, they stay white instead of turning green. Only the light of the sun can make green leaves.

If you cut through a tulip bulb, you find something else; you find the real secret of the tulip! For near the centre of the bulb, under the skins, there is a tiny, new bulb that is actually next year's tulip. And when this year's tulip has gone, the next one is ready to come up; it is ready down there in the bulb.

So there is a green tulip plant above, and a "white" tulip-plant (the bulb) below, and inside the white plant, under the earth, there is another one – next year's tulip.

When the tulip begins to grow from its bulb in spring, it is "in a hurry," it can hardly wait to reach the flower stage. So it only forms simple, green leaves with parallel stripes.

Other plants form sepals, the little green leaves of the calyx, but the tulip has no time for such things. At first you may think there is a calyx, there are green leaves that form a bud, but these green leaves turn red or yellow, and open up and become the blossom. So you see, the tulip is a plant that has not yet learned the difference between green sepals and coloured petals.

The difference between holding the petals of a tulip, or the petals of a rose between your fingers, is that the tulip-petals are thick and fleshy – they feel like green leaves – and the rose-petals feel very fine in comparison. The tulip is in too much of a hurry, it is a hasty plant, and it cannot produce fine petals, like the petals of the rose.

The tulip also produces a kind of fruit, but that is also done quickly, just a dry capsule around the seeds. And all the juices that other plants put into their fruit, goes downwards, into the bulb below. So you could say that the bulb of the tulip is a kind of "fruit," but it is not a fruit ripened by the sun. And the real fruits, up in the light, are dry and without juice.

Now we come to a relative, a cousin of the tulip, it belongs to the same family as tulips, lilies, hyacinths, daffodils, but the proud beautiful lilies and tulips are rather ashamed of the poor fellow, they look down upon him, and they are quite wrong to do so! That poor cousin of tulips and lilies is the onion.

The onion has no nice flowers like the tulip does, the flowers of the onion are tiny, greenish things, and these tiny flowers do not have much scent, and no colour. The warmth and the fire, that makes colours and scent, goes into the bulb. The

16. THE TULIP

onion keeps its scent in the leaves and stalks – within the bulb of the onion itself – it does not allow the scent to rise to the flowers, but keeps it down, and so it becomes the well-known pungent odour that makes your eyes water when you cut an onion.

For us, however, the onion is a very useful fellow, not as beautiful as its proud relatives, but an honest, useful helper in the kitchen. The next time you smell an onion, keep in mind that its scent might have gone into flower and fruit, but below the ground, it could only become the sharp smell it is.

Tulip
Leaves: simple, parallel veins
Bulb: watery leaves below ground
Roots: fleshy, no branches
Flower: no calyx
Spring plant: rushes to bear blossom

17. Seeds and Cotyledons

We divided the flowering plants into two groups. The flowering plants are those that learn the wonderful art of bearing blossoms from the wisdom in the sunlight. Some of the blossom-bearing plants are like young children, what they learn is still simple, their leaves and blossoms are still simple.

The tulip, beautiful as it is, has simple leaves, that is, the veins are parallel. The petals are thick, fleshy and a bit clumsy, in comparison to the petals of a pansy, or a rose, which feel very fine. It is an early flower that grows from a bulb, it does not wait for the full power and warmth of the sun.

There is another difference between the two great plant groups, between the simpler plants and the more perfect plants. And to understand this difference we shall consider the growing seed.

If you look at a little seed, you can hardly tell what plant will grow from it, unless you already know what plant the seed came from. It is like a little book that you have not read, and unless somebody tells you what is in the book, you don't know. The sun and the earth, the water and the air, open the little book and show you what is in it.

Imagine that you know what kind of plant is hidden in that seed, and you want this plant to grow in your garden. First of all you must make sure that the little seed will have a chance to grow properly. If there are weeds or grass where you put the seed into the earth, then the tiny plant that will come from the seed will not have a chance to grow properly; weeds or grass must be cleared away.

You then have to do a little digging. You must make the soil loose around the seed, and if it is too hard, the seed will

17. SEEDS AND COTYLEDONS

not grow. With some children, it takes quite a bit of digging before their minds are ready to receive the seed. Another thing to keep in mind is to choose the right time for sowing the seed. The different flowers in your garden have different times that are best for sowing. Of course, spring flowers must be sown earlier than autumn flowers. And the gardener knows the best time for sowing each kind of flower.

For thousands of years, people have been sowing seeds for flowers, and for crops, ever since the time of ancient Persia. And they found in those ancient times, that the moon has something to do with the growth of plants. Not only the sun and the earth and the water and the air, but the moon also has something to do with the growth of plants.

What they found was this: if you plant a seed just three days before the full moon, the plant will grow stronger, better, and faster, than if you plant it at any other time. The best time for a seed to be planted is three days before the full moon, when the moon itself is still "waxing" (that is "growing").

In our time, the knowledge that the waxing moon helps growing plants, has been forgotten – many people don't know about it, or don't believe in it. But those who do know it, and plant their seeds at the right time, they do get better results; their plants do grow faster and stronger.

The seed has been planted in the earth, it has been watered properly, the sun has been shining upon the earth, and the first thing to happen is that little seed begins to swell, just like a sponge swells, and gets bigger when you fill it with water. Then, from the seed tiny roots grow downwards into the earth and tiny leaves grow upwards towards the light.

The little leaves that come first, the leaves which come from the seed, are not at all like the real leaves that come later; they are either egg-shaped or heart-shaped. You know that in olden times when a king came into a city, heralds went before him. Heralds are men who announce to the people: "The king is coming, the king is coming!" And the king would only come once the heralds had announced him; they cleared the way for him. The little leaves that come from the seed, the seed-leaves, are like these heralds. They break through the earth and prepare the way for the true leaves, and for the flower that comes later. In botany, these seed-leaves or "herald-leaves" are called cotyledons. The true leaves grow from the stalk, not from the seed.

Now we come to the two groups of plants that you have heard so often: the flowering plants with leaves of parallel veins, and the flowering plants whose leaves have a network of veins. They also have differences in terms of these herald-leaves. The plants which have simple leaves, parallel veins, they only have one herald, their seed sends out only *one* cotyledon (monocotyledons in botany). The more perfect plants, the plants whose true leaves have a network of veins, their seeds send out two heralds, they have two cotyledons (dicotyledons).

18. The Rose

There are so many types of flowers and blossom-bearing plants that we could spend the whole term talking about the different kinds, and still we could not cover all of them.

We have learned about the tulip because it is a beautiful flower, which everybody knows and everybody has seen. The tulip belongs to one group of flowering plants (the group which is like younger children), the group that has parallel-veined leaves, and has only one seed-leaf or cotyledon.

From the other group – the more perfect plants, with network leaves, and two seed-leaves – we will look at another beautiful flower that everybody knows. We will look at the rose.

People have always called the rose the "queen" of all flowers, for no other plant has such beautiful petals, and no other plant has such a sweet scent. Why do we say about some people: they have "rosy" cheeks, their cheeks are "like a rose"? What is it that gives "rosy" cheeks? It is the blood flowing under the skin that gives rosy cheeks. When we run, our blood flows faster and we all have "rosy" cheeks, for our own blood is the colour of a red rose.

The rose plant is not as hasty as the tulip; it takes its time to bloom. The roses don't appear in spring; they appear in summer when the sun is at its height, when sun's light and sun's warmth are at their strongest.

The rose plant also takes great care over its green leaves. Each leaf has a "network of veins," and each leaf has an edge of tiny teeth, and these small leaves are set on a common stem so that, together, they are really one large leaf. You will often find seven smaller leaves growing from a common stem, making one larger leaf. The rose's leaves are quite different from the simple, straight-edged leaves of the tulip!

The rose not only loves the sun, it also loves the earth just as much as it loves the sun. The rose grows hard, strong roots deep down into the earth, and each root has branches, and these branches divide into smaller branches. It really is a proper little "tree" that grows from the rose plant into the earth, but a tree that grows upside down. Think of how different this is from the tulip, which only has straight, thin roots growing from the bulb. The tulip is in such haste to reach the flower that it is not quite at home on the earth; it just touches the earth. The tulip is like a person who runs on tiptoe, but the rose is like someone who walks with a firm, strong step.

It is not only by the roots that you can see that the rose loves the earth. You remember that the wood of trees is really earth, and that the trunk of a tree is really like earth reaching upwards. The rose is a "wood-plant." It draws the earth upwards and has slender wooden trunks and branches, as a result. It is a shrub, as we call all wood-plants that are not proper trees.

The branches of a rose plant do not just shoot up towards the light, they bow, they form an arc and the tip turns down slightly towards the earth again. That is quite different from the tulip; the tulip stalk has no time to form strong wood, the tulip has no wood at all, it does not care for earth, and the stalk goes straight up, away from the earth towards the light.

So you see, the rose does show that it loves the earth, and in its flower you can see that the rose is also a loving child of the sun. The tulip gets the calyx and the flower mixed up – its calyx becomes the flower. That never happens with the rose. The calyx leaves, the green sepals, are quite different from the petals – they close together and form a shell, like the chrysalis of the butterfly, and in that shell the rose petals are born, as the butterfly is born in the chrysalis.

The tulip, and all flowers which belong to that family: lily, hyacinth, narcissus, crocus, all those which do not care so

18. THE ROSE

The sepals of a rose (numbered from below up, from the stem towards the blossom). The first two have "whiskers" on both sides, the third only on one side, and the fourth and fifth have no "whiskers."

much for the earth, they follow the six-pointed star. But the rose loves the earth, and it follows the five-pointed star. The wild rose – the briar-rose – has five petals. The garden roses have much more than five petals, but if you count the petals of a garden rose, you will find that its number can always be divided by five. The rose knows the five-times-table!

When the rose petals have fallen off, and only the five sepals, the calyx, remain, it becomes a five-pointed star again. These five sepals have a secret, and I will tell you what this secret is. Some of the sepals that remain once the petals have gone – and some, but not all, of these sepals have little "whiskers."

Number one and number two have whiskers on both sides, and number three only has one whiskered side. Number four and number five do not have whiskers. You have now gone round the five-pointed star.

So, the rose that loves sun and earth, that is faithful to sun and earth, follows the five-pointed star; the tulips and lilies follow the six-pointed star.

19. The Rose Family

The rose is the plant that loves the earth as much as it loves the sun, and so, in the rose, the forces of the sun and the forces of the earth are in perfect harmony. Because sun and earth are balanced in the rose, it is the most perfect of all plants. Other plants may have more colourful petals, and others may have a stronger scent, but this perfect harmony of sun-forces and earth-forces exist in no other plant.

Think back to the people of ancient India who did not really care much for life on earth. The holy men of India did not work hard, they spent their time in prayer and their minds were so concentrated on heaven, that hunger and thirst, cold and heat, life on earth itself, meant very little to them. In the tulip, in the white lily, and in the narcissus, that only have thin roots, without wood, you can picture life in ancient India.

But it was different when you learned about Persia. Remember how the sun god, Ahura Mazda, showed King Jemshid a golden dagger in a dream, and the King understood that he had to make a fine plough. The Persians loved the sun, and they loved their sun-god, Ahura Mazda, but they also loved the earth. They were the first farmers, the first people to plant crops and flowers and fruit, and they changed the earth to make it a home for human beings.

You can easily see that the Persians were like the rose: they both love the sun and the earth. And the amazing thing is this: it was the Persians who changed the wild roses with only five petals, into the garden roses with many petals. And no one knows how they did it; no one today has the wisdom to change a briar-rose so that it becomes a garden-rose. You can "graft" a garden-rose onto the young stem of a wild rose, and it will

19. THE ROSE FAMILY

become a garden rose, but you must have a garden-rose plant first.

The rose is only one plant that belongs to a great family. All our fruit trees belong to the same family, to the rose family – apples, pears, cherries, plums, peaches, and apricots, they all belong to the rose family, and they all have blossoms with five petals. But with the rose, all the strength of the plant goes into the flower, and so, the fruits of the wild roses are the little red "hips" you can see in autumn. These rose-hips are not very juicy, but people used to make a sweet syrup from their juice – they needed an awful lot of hips to get enough juice from them. The fruit trees do not give all their strength to the flower, however, they give their strength to the fruit. Just think of the lovely smell there is when you bake apples! The apple tree sends its strength into the fruit, not into the blossom, and the scent is also given to the fruit.

So you see, the apple tree is a kind of rose which says: "I am not going to please human eyes with my blossoms, I am not going to please human noses with the scent of my petals, I am going to do more for them; I am going to please them with the refreshing taste of my fruits' juice."

To make the fruit, the strength of the earth is needed (you remember the ovary is a little earth) and so the apple plant needs more strength from the earth than the rose plant. That's why the apple plant must become a tree, it needs more wood, and that means more earth than the rose plant needs. The rose, which gives more strength to the lovely flower, can remain a shrub, with slender wooden stems, but only hips can grow on it. The apple, and all the other fruit trees, must have more help from the earth, and so they have to be trees.

All the members of the rose family – the garden rose, the apple tree, and the peach tree, and so on – all give us delight and joy in their different ways. And just as the rose came from Persia, these fruit trees have also come to us from Persia, from that ancient land when people loved the earth as well as the sun.

There is yet another important difference between apple blossoms and rose blossoms. The apple tree wants to have a juicy fruit, and therefore, has plenty of nectar in its blossoms – that is the sweet juice the bees seek to make honey. And so, when the apple tree is in blossom, you can always hear the hum of the many bees that come to collect the nectar. The rose is a "drier" plant, the hips are not very juicy, and the rose-blossom has no nectar for the bees. The bees do come to the rose to collect some pollen, as they also need pollen for their food, but you will never see as many bees around a rosebush as you will see around the apple tree. So, the rose gives all its strength to beauty, whereas, the apple and the other fruit trees give most of their strength to produce their juicy fruits.

20. The Cabbage

The difference between the apple plant and the rose plant is that the apple plant sends all its strength into the fruit, but the rose plant keeps its strength for the blossom. Plants do not need to give the same strength to every part; they can make one part stronger and other parts smaller or weaker, for example, rhubarb – all its strength goes into the stalks and leaves, and all the cactus' strength goes into the stalk.

If you play three notes on the piano, you can make one louder, and the other two softer, or one note longer, and the others shorter. There is one plant that is a real artist when it comes to making one part louder, that is bigger and stronger than other parts, and then sometimes making others bigger and stronger. As it happens, this plant has been helped to play so many different tunes on its parts, by human beings, by gardeners and farmers. This plant would not have so many different shapes without human help, but it also needed to be a special plant in the first place so that gardeners could work on it. This plant has shapes that are so different from one another, that you would never guess, that to begin with, it was only one plant. This plant can disguise itself in so many ways that you would never think it is really the same plant – it gives its strength to a different part all the time. This plant does not follow the five-pointed star, or the six-pointed star; instead it follows the four-pointed star. Its leaves stand in a cross.

First, let us see again what the parts of this plant are:
1. Flower
2. Stem
3. Leaves
4. Roots

This plant can either send all its strength into the flower, or into the stem, or into the leaves, or into the roots, and each time you get something quite different. But remember one thing: when the plant was evolving, once it had put all its strength into the root, the new plant with a great thick root, stayed like this; it became a separate plant in its own right. When this plant – with human help – gave all its strength to the root, the result was a new plant, which we call a turnip. The turnip is one disguise of this special plant. In the disguise of the turnip all the strength goes into the root.

But farmers and gardeners also made this plant give all its strength to the stem, and the stem grew fat and full and thick, instead of the root, and the result was another vegetable. It is not so well-known in Britain, but on the Continent it is used as much as the turnip. This new vegetable with all the strength going into the stem is the kohlrabi.

The next disguise you all know. When all the strength goes into the green leaves it becomes the cabbage. The cabbage and the turnip both come from the same special plant originally.

Now the next disguise would be one in which all the strength goes into the flower; but gardeners break the large green leaves and cover the large bud, so that the sun cannot make a flower, and so only the "receptacle" grows enormously. However, in the receptacle you can see the many flower-buds. The receptacle is white, because the sun cannot reach it, and it is soft. This is, of course, the cauliflower. Its real name should be cabbage-flower, for it is a kind of cabbage, which gives all its strength to the flowers, or in any case to the receptacle, which is part of the flower.

There is still one more disguise. Have you noticed that in places where leaves first begin to grow, plants show a little spot called an "eye" because it is shaped like an eye. This special plant can also put all its strength into these eyes – the little leaf beginnings – and a tiny cabbage grows from each eye. These tiny cabbages are called brussel-sprouts.

20. THE CABBAGE

So there are five different vegetables, all coming from the same plant, which grows wild in warmer countries and is called wild cabbage. All these vegetables are "disguises" of the wild cabbage, but whether it is cabbage, or cauliflower or turnip, depends on where the plant puts its strength – into leaf or blossom or root.

21. The Nettle

You have heard of the flowers and plants that we grow in our gardens for their beauty, like tulips and roses, or plants we grow because we eat them, like the different kinds of cabbages. But now we are going to learn about a plant that few people like; a plant that does not have beautiful flowers or juicy fruit, nor does it have a pleasant smell, and yet it really is a nice plant. It is the stinging nettle.

The first thing to learn about the stinging nettle is that it is a plant that has great strength. You see, the plants in our gardens need a lot of protection – they must have the right soil, they must be protected from weeds, they must be watered if there is not enough rain. Our garden plants all need a helping hand from us, but the nettle is a wild plant and it can grow in any kind of soil; it can even grow in the worst soil.

The nettle is not only a strong plant; if you look at it carefully you could even say it is beautiful. Just look at the beautiful order of the leaves. They are in four rows, and the two leaves of the bottom row stand in a cross formation with the two leaves of the second row, the third row in a cross with the second, and the fourth in a cross with the third. And each leaf is beautiful in its own right, with fine-feathered edges ending in a point.

But these beautiful leaves have little "hairs," and it is these hairs that sting you. These "hairs" are not really hair; they are really stiff bristles, rather like fine glass spikes. When you touch these bristles they break, the broken bits penetrate the skin, but the juice that was inside the bristle is the thing that makes your skin feel as if it were burnt, that is what stings you.

The nettles that grow in our part of the world only sting,

and the sting soon passes without doing any real harm. But in hot countries, in the tropics, there are nettles whose sting is much worse and it can make you ill for days. So our nettle is really a nice fellow, he doesn't do any real harm.

The nettle does grow blossoms, pale light purple blossoms, but they are small and have no scent, and these blossoms don't attract any insects. The nettle has its pollen carried away by the wind. The nettle is, however, different in one way from the other plants you have learnt about. Some nettles have flowers with stamens only, no pistil and ovary, and others have flowers with pistil and ovary, but no stamens. So, the wind brings the pollen of the stamen to the nettle blossoms that have only pistil and ovary.

The nettle does not attract bees or butterflies with its flowers, yet it is a great friend of some butterflies. These butterflies – especially the red admiral – lay their eggs upon nettle leaves, and the caterpillars that come from the eggs feed on the nettle leaves; they feed on the leaves that sting us, but they don't seem to sting the caterpillars. And from these caterpillars come the beautiful red admiral butterflies. When the gardeners use weedkillers to get rid of nettles, unfortunately, they also get rid of the butterflies. The nettle does not have beautiful blossoms of its own, but it is kind to the caterpillars, and, one could say, produces the "flying blossoms," the butterflies.

In spring, the young nettle leaves can be gathered and cooked in a soup, and in the old days when people knew more about the healing powers of plants, they used to give nettle soup to invalids, to help them get strong again. It worked very well, and even works today if people bother to pick the young leaves. In addition, the fibres of the nettle were used to make a strong yarn to weave a strong, warm, kind of cloth.

So, the nettle is actually a good fellow, it is rather like people who are rough and uncouth on the outside, but have a really good and kind heart inside, and will come and help you when you are in need.

The Plants We Use

22. The Oak

You have heard in geography how important trees are. In the Scandinavian countries the forests are a source of wealth for the people, because they provide timber for building, for furniture, for packing cases, for paper, and for matchsticks – whole forests are often cut down for this purpose. But you have also heard that trees and forests are important in other ways too, and where people have cut down forests to use the wood without re-planting, they have paid a terrible price for it. Huge areas of South America, Africa, and Asia would not be barren; they would have fertile soil if people had not cut down all the trees. Without tree roots to hold the precious soil in place, the rains washed the soil away, and most of it was swept into the sea via streams and rivers. So the trees are the protectors of the soil. But the trees are not only protectors of the soil, they help to make new soil by shedding their leaves. Over time, tree leaves become wonderful soil as they decay.

In ancient times, for example, when the wise Celtic Druids lived, people revered the trees of the forest; they held them in awe. They felt no temple with marble pillars could be as beautiful as the trunks of forest trees reaching up from the ground, and the "vault" of green leaves and branches above. That is why they worshiped in forest-glades, in the great forests that once covered Britain, of which little is left now. There was one forest tree that the Druids treated with great reverence, the oak tree. The name of this wise people comes from the oak tree, for *drus* or *drys*, was their name for the oak tree (in Greek, oak is *drys* too) and a "Druid" means a man who is like the oak.

The oak tree is certainly a picture of strength. The wood is very hard, the branches are knobbly and thick and the foliage is a deep green. Think how different the slender trunk of a birch tree looks compared with the might and strength of the oak tree. The birch is a tree that looks loveliest when it is young, but the oak tree gets more beautiful, more majestic, the older it is. And the oak can get very old indeed, easily two hundred years. In Allonville, Normandy, is an oak tree that is sixty feet high and thirty feet thick, which is one thousand years old.

But the oak does not form its mighty trunk, and its hard, strong wood quickly. The oak is a slow-growing tree, like a person who does not believe in doing things quickly and carelessly, but a person who wants to do everything well and carefully, a person who works steadily, but slowly and thoroughly.

In the spring, when all other trees are already dressed in their new green foliage, the branches of the oak tree are still bare; it takes its time and its leaves come out later than the others do. But the oak also holds its leaves for longer than most other trees. In autumn, when all other trees have lost their leaves, the oak tree still holds onto its leaves, even though they are already brown and dried up, and you can hear them rustle in the wind.

By the time the oak tree's green leaves appear, it is so late in spring that its blossoms appear at the same time. But an oak tree waits for thirty years before any blossoms grow upon it. For the first thirty years of its life an oak only has leaves and no blossoms.

A fellow like the oak tree that wants to be strong and firm, could not very well be expected to have blossoms with beautiful coloured petals; that kind of blossom would not go with the oak at all. It would be like expecting an eagle to sing sweetly like a blackbird, or to twitter like a sparrow. No, the blossoms of the oak tree are tiny catkins, and you hardly notice them at all. The oak tree cannot produce sweet fruit either, that would not be in style for a tree of such strength. The fruit of the oak tree is the acorn in their little cups.

Now acorns, little as they are, are not the kind of seeds that can be scattered by the wind, they fall down under their tree.

22. THE OAK

And from all these acorns under the oak tree not a single new tree would grow, for a seed needs sunlight to grow even after it has sprouted. The foliage of the old tree lets a little sunlight through, but not enough to allow a new tree to grow under it.

So how do new oak trees grow? A certain large bird with light brown and white feathers, and black and white wings and tail, likes acorns. It is called a jay. The jay wants to eat as many as he can find in autumn. There are plenty of acorns, and the jay can't eat all of them, but he does not want to leave them either, so he takes lots of acorns away, and when he thinks he has found the right spot, he digs a hole in the ground with his beak, puts an acorn into the hole, and covers it with earth. He does this with all the acorns he finds. Later he comes back to dig up the acorns that he has put into the ground, but he always misses some, and next spring, here and there, the tiny shoots of new oak trees come up through the ground. So the jay helps to make new oak trees grow.

A tough, strong tree like the oak cannot grow pretty blossoms or sweet fruit, but there are also occasions when the oak tree shows that it has sweetness. In the same way you sometimes meet a person who seems tough and rough, but under that roughness, when you get to know this person better, there is a gentle heart, which only shows itself on special occasions.

There is a particular wasp called the gall wasp. This wasp makes a hole in the oak leaves with its sting, and it lays its eggs in these holes, and then flies away. It leaves its children to the care of the oak tree, and this strong, hard tree becomes a gentle protector to the tiny wormlike larvae, which come from the eggs. Around each little larva, the leaf grows a little hollow ball the size of a cherry, and you can often see these balls (gall-apples) on oak leaves. Inside these little green balls there is a sweet juice, and the tiny insect eats that juice and lives on it, until it becomes a wasp and can fly away. So you could say the oak has two kinds of fruit: the acorn, and the little gall-apples that are only grown for the larvae of the gall-wasp. The little balls, the gall-apples have a sweet juice, and also a very bitter juice within them, and this other juice is used for tanning leather. Animal skins are kept soft and supple by tanning in the juice of gall-apples.

The Druids were quite right when they thought there was something special about the oak tree. For itself, its only fruit is the acorn, but for the little creatures, it grows the gall apple with sweet juice inside.

23. The Birch

You could call the oak tree the king of the woods, because it is the tree of might and power and strength. And the birch tree could rightly be called the queen of the woods, because it is the most elegant and graceful of our trees. Let us compare the trunk of a birch tree with the trunk of an oak tree. With the oak tree you can see immediately that it has a trunk meant to resist the wind, to stand up against the wind. In comparison, the slender and supple trunk of the birch tree is meant to sway with the wind; the birch tree is friends with the wind, it sways and dances with the wind. And while the bark of the oak tree is rough and dark, the bark of the birch tree is smooth and white with black strips.

The leaves on an oak tree with their simple round forms also look quite clumsy compared with the triangular birch leaves with tiny, pointed teeth on their edges. Birch leaves are a very light green, and they flutter in the wind like little flags. In autumn, the birch leaves are shiny yellow, like gold.

If you look at great strong trees like chestnut and oak, they have a proper crown. The crown of a tree is the part that rises above the trunk, and it is formed when strong branches grow out from the trunk, and little twigs then grow on these great branches. But it is not so with the birch. On a birch tree the thin, bending branches grow straight out from the trunk; it is as if the birch says: "I don't want strong branches which stand up against the wind, I want thin, supple limbs which move and sway with the wind," and the birch, therefore, has no proper crown.

The birch is really a youthful tree, a tree that wants to be young, and it does not grow very old like the oak tree does.

The birch usually lives for about one hundred years, but not much longer. The birch tree has one thing in common with the oak tree – it does not have pretty blossoms, with coloured petals. The blossoms of the birch tree are also tiny and stand in catkins. These catkins hang down from the branches like streamers in spring. But the seeds that grow from the catkins are quite different from the heavy plump acorns of the oak. The birch seeds are so tiny, that you can hardly see them and you have to look very closely, or use a magnifying glass, to see that each seed has tiny wings, and these tiny, winged seeds are easily carried far and wide by the wind. So you see, the wind and the birch tree are real friends, and the birch tree does not need birds to get its seed scattered; its friend, the wind, will carry the winged tiny seeds through the air.

Now the tiny blossoms in the catkins don't have a sweet smell, and the tiny seeds don't have a sweet taste, so it seems rather disappointing that such a friendly, lovely tree like the birch has no sweetness. There is sweetness, however. If you make a little hole in the bark of a birch tree in May and stick a tube or a straw into the hole, little drops of sap will soon flow from the tree, and that sap has a lovely sweet taste. You see the birch does not send its sweetness into blossoms or seeds; it keeps the sweetness under the bark and sends it to its twigs and leaves.

This sweet sap of the birch tree makes a very good tonic. If people are tired and worn out, this birch tonic gives them new strength. And this is not surprising, because the birch is a tree of youth, and so, its juice can make elderly, tired people feel younger and fresher.

The birch is not only a youthful but also a useful tree. The wood of the birch tree is soft, but this soft wood is sawn into thin layers which are glued together to make plywood which is strong and flexible. There are great sawmills and factories in Finland that benefit from birch wood, and the silver birch is also the national tree.

The bark of the birch tree is so wonderfully smooth. It

23. THE BIRCH

keeps out water and the natives of North America knew this, and made their canoes from the bark of birch trees. Their birch boats weighed very little and could easily be carried from one river to another. The wood is also used for furniture and flooring.

This beautiful tree that has so many uses demands very little for itself. It grows in poor soil, and it can stand a harsh climate and cold winters with snow and ice much better that the proud oak tree. That is why you find birch trees, whole birch forests, in Scandinavia, in the far north of Norway, Sweden, and Finland where no oak trees would grow. There are thick woods, whole forests of birch, and the slender trees grow so close together that it is very hard to get through. But even the birch tree cannot grow in the very far north. Above the Arctic Circle you would no longer find the long, slender birch tree, but a pretty little bush, no more than knee-height, with leaves about one centimetre (half an inch) big. This little bush is also a kind of birch, called dwarf-birch. In spring, when tiny catkins hang from the branches you would recognize that this is a little sister of our queen of the woods, the birch tree.

The birch tree which asks so little for itself – it does not ask for good soil and it does not ask for a lot of warm sunshine – is like a person who is beautiful and kind, but does not make any fuss about it, does not draw attention to themselves, and is humble, modest and content with very little.

24. The Palm Tree

The root of a plant belongs to the earth, and it loves the earth, it grows downwards. Roots will always grow away from light because they like the darkness; they like to feel covered by Mother Earth. But the blossom turns towards the light, it opens its petals to the light, and the petals get their lovely colours and their scent from the light. The green leaves and the stem, they stand in between: as the colour green stands between bright yellow and the dark blue in your paint-pots, and is a mixture of both; so the green leaves and the stem stand between the sun-loving blossom and the earth-loving roots. The whole plant belongs to both, to sun and to earth. The earth is the mother and the sun is the father of the plants. In the blossom, the plant shows its love for Father Sun, and in the root, its love for Mother Earth; in the green leaves and stem, it shows its love for both together.

But the plants in the world are not all the same. In hot countries, where the warmth and light of the sun is very strong, the blossoms get big and colourful, and the whole plant wants to reach towards the sun; the leaves get large, the stems grow tall, and the tree trunks grow higher and higher. But the further north you go, the more the plants shrink – the blossoms get smaller, stems and tree trunks get shorter. In the far north, in the Arctic, the birch tree becomes a dwarf-birch, a little dwarf, a tiny birch that clings to the earth.

So one could say that the whole plant-world becomes more like a blossom the nearer we come to the tropical lands of the Equator, and more like a root the nearer we come to the cold lands in the north. When we have this in mind we can under-

24. THE PALM TREE

stand trees of the tropical lands better, trees such as the palm trees.

A palm tree looks so different from our trees because the trunk shoots straight up without any branches, there is no spreading out from the trunk. A palm goes straight up towards the light and the leaves are on top. But all the strength of the tree that has not been used for branches and twigs now goes into the leaves, and that is why the leaves of a palm tree get so big, because the strength that other trees use for branches goes into these leaves. They are big leaves indeed: the leaves of the coconut palm reach a length of five metres (fifteen feet), more than twice as tall as a man. But they are so high up that they don't look as large as they really are. They are high up because a coconut tree can grow to a height of thirty metres (a hundred feet). These large leaves are not just broad leaves, they are feathered, which means that they look like ferns.

There are many kinds of palm trees, particular to different regions and climates. The coconut palm likes the coast where damp air from the sea blows through its leaves. Its fruit, the coconut, sometimes falls into the sea. And as it can wait six months before it begins to sprout, the sea often carries it to another island or coast before it has sprouted. Coconut palms, therefore, can be found anywhere in warm coastal regions of Africa, India, and South America.

Now when you take a fruit that we are familiar with – an apple for instance – you find the seeds within the core, or in a cherry, the seed is inside the stone. With the coconut, which is a big fruit, you have to look more carefully to find the seed. On the hard shell you will see three dots, and when you crack the shell, you see a brown skin. Under the skin is the white "flesh" of the coconut and inside the flesh is the "milk" as it is known. But the flesh and the milk are only there to feed the seed until it is big enough and strong enough to surface and grow into a tall, strong palm tree. The seed is quite small at first – it looks like a small, soft fir-cone and you will find it underneath the shell where the three dots are. This little cone will slowly absorb the milk and the flesh, and grow bigger and stronger, developing tiny green leaves. It has the strength to begin to grow into a tall tree then.

We think of coconuts as a food, something to eat. But to the people of Africa and India the coconut has many more uses. When a coconut is first taken from the tree, it has a smooth outer skin, either green or brown. Within this skin you find rough, fuzzy hair known as "coir," and within the coir, you find the hard shell. This rough hair is used to make mats and nets, and the large leaves are used to make roofs on huts, the "feathers," the strips of the leaves, are woven into baskets, and the long middle ribs are twisted together to make strong ropes. The young shoots of new leaves make a tasty vegetable, and the sap of the palm tree, if it is left standing for some months, becomes sweet palm wine. The wood is of course used for building.

There is an even more important use for the coconut. In the tropical lands of Africa, India, South America, and especially in the islands of the Pacific Ocean, there are large plantations of coconut palms. In these plantations the white flesh of the coconut is taken and dried in the sun. This dried flesh is called copra, and it does not smell nice, it smells like oil that has gone rancid. In the white, dried copra there is oil. The copra is put into a press, and the oil pressed out. The finer oil that comes from the copra was once used to make margarine, and the less fine oil is used for making soap and cosmetics. The soaps we use often contain oil that comes from the coconut. So this tall, tropical coconut tree is important to us.

25. Tea, Sugar and Coffee

Not only do we know the coconut tree from far away tropical lands, but there are also other plants which grow in foreign lands, and which we use here in our daily lives. Let us begin with the beverage nearly everybody drinks in the morning: tea.

You know tea only as little dried leaves and from these, you could hardly imagine what the plant they came from really looks like. Tea is really a tree, but the Chinese, the first people to discover that the leaves made a nice drink, used to prune the young stems of this tree down, so that they didn't grow higher than about a metre (three feet). A tea plantation is therefore a large field planted with shrubs. The tea shrubs are planted just over a metre (about four feet) apart. One could say that the tea plant is a cousin, a relative, of a flower you sometimes see in gardens here, the camellia, which can be a white or red flower. And like the camellia, the tea plant bears white or pink, sweet-scented blossoms that later become small round fruits. But tea is not grown for its petals or for its fruits, but for its leaves. Not all the leaves are used for tea, however, only the tender little leaves on the tips of young shoots. These little leaves are often picked by women. In a large plantation there may be five or six hundred women who go through the rows of shrubs with a large basket on their backs, pick the newly grown leaves and throw them into the basket. The little green leaves from all these hundreds of big baskets are then rolled, dried in special sheds, and then they are fermented, a process that gives the tea its special taste and smell. It is at this stage of fermentation that the leaves turn black or dark brown.

The first people to find out that tea had a stimulating effect

– it makes you more awake, and also more talkative – were the Chinese. About one thousand years ago, it was only the Emperor of China, and the wise men of his court that drank tea, for they thought tea should stimulate only clever talk, not chattering. But the tea usage soon spread to all the people in China.

Why does tea have the effect of making people more awake and talkative?

Because the leaves contain a drug, a weak poison, but it is poison all the same. It has a stimulating effect and that is why people don't mind the very weak poison in it, because a healthy body can cope with a little poison. Nowadays, the tea we drink comes not only from China but from Sri Lanka, Indonesia, Kenya, and from Darjeeling in India. So when you drink tea, think of the women who work with bent backs all day, amongst the long rows of shrubs, in the plantations.

The tea we drink comes from a plant grown in the east, in Asia. But the sugar we put into the tea comes from the opposite side of the world, from the west, from the island of Jamaica, which lies off the coast of America. Most of you will have seen reeds or rushes growing on the banks of rivers, lakes or canals, with long hollow stems and stiff, pointed leaves. And sugar cane, as it is called, looks rather like these rushes.

Now we come to something very interesting. Nearly every plant produces sweetness, or sugar. The tiny drops of honey that the bees collect from the blossoms are also a kind of sugar. The sweet taste of fruit, apples, oranges, and bananas, comes from the sugar they contain. Some plants produce sugar in their blossoms (nectar) or in their fruits, and this sugar that comes from blossom or fruit is best for the human body; it is the sugar we can digest best. It has a special name: glucose. Fructose or "fruit sugar" is also found in fruit.

Other plants don't let their sugar reach the blossom, where the sunlight can work on it, and give the sugar its blessing. They keep their sugar further down, in the stem, in the leaves, like oak and birch. Sugar cane also does this – it keeps its sweetness in the stem, it does not go into blossoms or fruit. You could say it is a more selfish plant; it keeps the sugar to itself. In blossoms and fruit, the plant gives the sugar away. This second kind of sugar, like cane sugar, is not so easily digested and not so good for us as fruit sugar. Then there are plants that keep their sugar down in the roots; the sweet taste of carrots comes from the sugar they have. And the sugar beet, which is also used to make sugar, is a root – a bit like the beetroot – that is full of sugar. This sugar made from roots, from sugar beet, is the one that is hardest to digest and worse for us than any of the others.

But fruits do not have a lot of sugar, not enough for all the sugar needed in the world, and sugar made from sugar cane, as we have it, is the next best thing. But there are many countries that only have beet-sugar, sugar that comes from roots.

The sugar-cane in Jamaica is also grown on plantations. It grows up to a height of six or seven metres (twenty feet), more than three times the size of a man. And the tall plants are planted so close together that you could not walk between them. It is like a forest of high grass or high rushes, and the leaves between the cane stalks make it impossible to walk through. When the canes are high enough, a strange thing is done to the plantation: it is set on fire. Not everything is burnt to ashes, however, for when the flames die down, only the leaves have gone, the main stem remains. The workers come at this point – usually strong men – with long bush-knives like short swords, and they cut the long stems down. These long stems, the canes, as they are called, are taken to factories where they are crushed between heavy rollers, and a sweet, yellow, pungent-smelling liquid is the result. This is the sweet juice of the cane, which is then boiled to become a thick syrup. A machine extracts the sugar-crystals from the syrup and what is left is known as molasses.

The crystal-sugar that is extracted is yellowish brown, and this is the natural colour of sugar. Demerara sugar, for example, keeps its natural colour. Chemicals are used to make white sugar, so brown sugar is the most natural sugar.

Although the sugar we use is grown in Jamaica, this is not actually the home of the sugar plant. The plant was first grown and used in India, so when you put sugar into your tea you bring together two plants whose origins are in India, in Asia.

Coffee comes from a small tree that grows in parts of Africa, Arabia, India, and Central America. Coffee trees have fruits that look like red berries. Inside the berries are two seeds, and these seeds are the coffee beans. They are green and get their coffee-brown colour, as well as their taste, once they are roasted.

Another plant of great importance to us, but not for food, is cotton. It grows in hot countries – India, Africa, America, for example. It is a shrub with long leaves and is a cousin of the hollyhock which grows in Britain. It has red or yellow blossoms from which round green fruits called bolls apear. This fruit does not fall from the plant, however. When it is ripe it opens, and inside are seeds covered with long white hair. This white hair, five to eight centimetres (two to three inches) long, is the cotton that is so important to us.

From this plant we can make the cotton wool we put on wounds, as well as the cotton we use to make shirts, blouses, underclothes, sheets, tablecloths, and so on.

When you put on your dress or shirt in the morning, and when you take tea and sugar, you are using plants from the far corners of the world – seeds, stalks, leaves, and fruits.

26. Grass and Cereals

One can think of trees and flowers and shrubs and herbs as the green clothes of the earth, green clothes that cover the dark soil. But there is something that is not a tree or a flower or a shrub, yet it is, perhaps, the most wonderful plant of all, it is grass, the plant of the meadows and lawns and fields. There are great plains of grass around the world: the puszta of Hungary, the prairie of America, the savannah in Africa, and the steppes of Asia. The grass of our meadows and lawns, of the fields and of the wide plains, is like a fur that covers the skin of an animal.

The light of the sun, as you may have learned in your physics lessons, only moves in straight lines. Think of a sunray coming to earth from above – and the earth answers by sending up a straight blade of grass.

Other plants grow upwards (vertically), as well as sideways (horizontally), and their leaves stand or hang, flat and broad. So the strength of the other plants goes in two directions, upwards in trunk and stem, as well as sideways, in branches and leaves. But grass is a plant in which everything strives to move upwards; it is as though the whole plant wants to be a stem, and nothing else. Grass does have a stem as well as leaves, but the leaves don't stand on little stalks on either side of the stem. The leaves of grass grow on the stem and are rolled round the stem, like a sheath, and only the last part, the top of the leaf, moves away from the stem.

A water lily, for instance, is the very opposite of grass. A water lily with its flat, round leaves, floating on the water, is a plant that has no vertical strength at all; it is a plant that wants to be horizontal. But grass is like a spear pointing upwards, a magic spear, because it is a spear that can stand upright by its own strength.

Grass is a very strong plant – it does not stop growing as trees and most wild plants do in winter (trees do not grow in winter – you can see it in the "rings" when a trunk is cut). But grass grows all the year, though more slowly in winter, that is why a lawn has to be continually mown. Think of the scent of freshly mowed grass, or the scent of freshly cut hay; in that wonderful scent you can smell its strength and vigour.

Other plants grow in both directions, vertically as well as horizontally. Grass wants to grow upwards only – the whole plant wants to be a stem. Another peculiar thing about grass is that it has no coloured flowers, no blossoms. Other plants have at the very least tiny blossoms, like the catkins of oak and birch and willow, but grass has no petals at all, only small dry husks. Where other plants have blossoms, the grass grows tiny green or brown husks containing little feathery stigmas, and stamens with fine filaments, but no blossoms like other plants have

Many other plants also have a fleshy fruit, and inside the fruit are the seeds, but grass only has the seeds – they are called grains – and they have no juicy fruit around them. There is something missing from a plant if it has no coloured blossom, no juicy fruit. It is as though the plant has given up something that other plants have, and all the strength that other plants use for beautiful flowers and round fruit, all that strength goes into the seed, into the grain, and it is hidden there.

The strength of grass does not show itself outwardly in flowers, nor does it show itself in fruit, it remains hidden and concentrated in the little grains, the seeds. It is as though the grass says: "I don't want to show myself in the beautiful dress of a flower, or in the sweet taste of a fruit; I give all my strength to my children, the little grains, so that they might become straight, upright beings like me."

The strength that the grass puts into its grains is of great importance to human beings. Many thousands of years ago, at the time of ancient Persia, or the new stone age (neolithic) as it is called, there were wise men that had the knowledge to grow the different kinds of corn – wheat, barley, oats, and rye – from some kinds of wild grass. No one knows today how it was done, but all the kinds of grain we use for bread, and for everything else made from flour, came from the first plants grown by these wise men, from wild grass. You call your breakfast food "cereal," but in botany, cereal is the name for all the grain-bearing plants that give us flour for our food. They are all different kinds of grass, however.

The concentrated strength of the plant is in all these the grains that are milled, that are ground by mills into flour. Each grain has the strength to become a new plant, and what very great strength it is, because wheat and barley do not have blossoms or fruit. That is why bread is so nourishing. When you eat bread, you eat the strength that could have made a new, straight, upright plant grow.

In different parts of the world, different grains and cereals are grown. Many people in Asia grow rice, in China, for example. Rice is also a kind of grass. In the west, the Native Americans had maize before the Europeans came who planted vast wheat fields in America. But maize, the American cereal, is also just another kind of grass. It is very interesting to see the difference between what is grown in Asia and America: in America, maize has a large cob and large round grains; and in Asia, tiny grains of rice grow. Our cereals – wheat, barley, and so on – stand in between.

You know that there is a great strength in grains of wheat, but in some bread that you can buy, the whole grain is not used. With white bread and white rolls especially, you get very little of the growing-strength of the grain. But if you get wholemeal flour, as it is called, and bake your own bread, you can almost taste the strength that earth and sun give to the grain.

If you think about how many things we eat need flour, and

if you consider that all the meat we eat comes from animals that must be fed on grass or hay, then you will realize that, of all the plants in the world, grass is the most important of all for us. Grass, the plant that is the most upright, grows straight up so that it can meet the sunrays from above.

27. Leaves and Blossoms

When you look at a plant, perhaps you like the flowers, or perhaps you think how useful or important the plant is for human beings. Perhaps you can feel the strength and power of the oak, the slender grace of the birch, or the modesty of grass, which denies itself pretty flowers to give all its upright strength to the seeds.

But you have to observe more carefully and look closer at plants to discover the wonderful wisdom that works within them. In some plants the leaves are long and come together at the base of the stem – the dandelion is such a plant. And when there are such long leaves meeting where the plant comes out of the earth, in the middle of the leaf, inside, there is always a groove. You will find that plants with long leaves like these, with inside grooves, always have long single roots that go deep down into the earth.

What is the connection between roots and leaves? When it rains, the leaves catch a lot of rain, but plants cannot absorb water with their leaves, they can only absorb it with their roots. In the grooves on the leaves, is a kind of "gutter" that guides the raindrops down to the root. The leaves of beetroots, radishes, and turnips are formed in this way, in order to guide rainwater down to the single root. The leaves form a kind of funnel for the root.

However, the leaves of an oak tree are different. They will shed the rain that falls on them anywhere under the oak tree. And this is appropriate for the oak tree, because the roots of the oak tree spread in all directions from the trunk, like the spokes of a wheel. If all the water from above were funneled to one middle root, then all the other roots would get nothing.

Leaves and roots, therefore, are always "suited" to each other. Of course, the oak tree itself has no brain, and neither does the turnip. It is not the plants themselves that have "thought out" which leaves or roots are best for them. It is the wisdom of God that works throughout nature and gives each root the right kind of leaf.

But now we come to an important question: why do plants have green leaves at all? To answer this question you only have to remember two things you know already: trees do not grow during the winter, but we also know that leaf trees have no leaves in winter, therefore the leaves must have something to do with growing. (The evergreens – fir and holly, for example – grow a little during the winter, but they grow more quickly during the summer.)

All plants grow because of their leaves. The leaves give food to the tree, and the sap carries the food from leaf to stem, trunk, and branch. Leaves can do what humans and animals cannot do: make food from air. From this food the plant builds its body and grows. If you stripped a tree in the summer of all its leaves it would not only stop growing, it would die.

So if you look at the green leaves of any plant, don't think that they just hang there for decoration. These green leaves are actually working. They collect food for the plant from the air with the help of sunlight. A little food is also drawn from the soil by the roots, but only very little. The roots provide the plant with water, and the leaves provide the food. The green leaves are the workers in the plant; it is through their work that the plant grows.

Now we come to the blossom. The blossom and what is inside the blossom have nothing to do with work; something different happens in the blossom. In the middle of the blossom there is something that looks like a tiny, green bottle, and in that little bottle are very tiny seeds. The lower part of the bottle is called the ovary. From the seeds in the ovary alone, however, nothing could grow.

27. LEAVES AND BLOSSOMS

You know the fairy story of the Sleeping Beauty who has to be awakened by a prince. Each little seed in the ovary is like a "Sleeping Beauty," and it must wait for a prince. Standing around the ovary are little stalks with a ball of yellow dust on their top. The yellow dust is called pollen and the pollen is the prince who can waken the seeds in the ovary. But the pollen can only get into the bottle – into the ovary – through the bottle neck, and if the flower is, a rose, for example, the pollen must come from another rose, because pollen and seed that grow inside the same blossom are no use. That is why the flowers need insects or wind to carry pollen from rose to rose, from tulip to tulip, or from violet to violet. The wind and insects are the horses of the fairy-prince.

When the pollen of one rose comes to the opening of the little bottle in another rose, travels down into the ovary and touches the seeds, they wake up. The whole ovary grows, and in time a new rose grows from the seeds.

What happens in the green leaves and what happens in the blossom, therefore, are two different things. The green leaves work for the plant in its current state – they feed the plant and everything on it, including the blossom. But the task of the blossom, the pollen, and the ovary inside the blossom, is to prepare for the future plant. Green leaves work and feed the current plant; pollen and ovary prepare the future plant, the plant that will grow next year. They do no work for the current plant. The coloured petals do not work either. With their colours and their nectar, they entice butterflies and bees and other insects to come, and carry pollen from flower to flower.

The leaves, the petals, the pollen, the ovary and seeds, are really one being; they are the plant. In the same way, your arms and legs do different things but are all part of you.

And now we come to the beehive. At first, it seems quite different to a plant – you would not dream of thinking it is like a plant, but it is. It is like a plant and unlike a plant at the same time.

In a beehive there are the worker bees that bring the food, the honey, to the hive, but there are also bees that do no work at all. The queen bee does not work; instead she lays eggs, tiny, tiny eggs. The queen bee is like the ovary, and the other bees that do no work – the drones – they are like the pollen. All the bees work together so that the hive is like one being, like one plant. Bees and plants are good friends and they help each other. A beehive can be compared to a flying plant, and a plant can be compared to bees that are stuck in one place.

28. Bees

It is easy to see why there is an attraction between bees and plants, because a hive of bees, is like one plant. In a plant, every leaf, root, petal, stamen, and ovary are just individual parts that make up the whole plant. If you take a leaf away from the tree, the leaf dies. Similarly, a single bee lives and works only for the hive as a whole, and if you take it away from the hive it will soon die. A finger from your hand is nothing by itself; it only has purpose when it is part of your hand. A leaf by itself is nothing; it only has a life and a purpose, when it is on a branch as part of the tree. A single bee by itself is also nothing; it is only a small part of the whole hive.

However, there are times when flowers send one part of themselves, their seeds, out into the world. The seeds must leave so that new plants can grow, and something similar also happens in the beehive.

One day in early summer, from the hive there comes a strange, excited humming, unlike the soft contented humming bees usually make. On this day, many bees are getting ready to move from the hive they know as home. There are about eighty thousand bees in the hive, and every day more baby bees appear from eggs, so the hive is getting too crowded. Many bees must, therefore, move out, and the excited humming means that they will move out that day; they will swarm.

You see a steady stream of bees rushing to the small opening at the bottom of the hive. They push and shove and climb over one other, and in a few seconds the air is filled with thousands of bees, darting and circling in the sunlight. The swarm draws closer together, and like a cloud driven by the wind, it drifts towards a nearby tree. Some of the bees must be extremely strong, for more and more bees come and cling to the bodies of these first ones, until the whole swarm, perhaps fifty thousand bees, hang from the tree like one body, one large humming mass.

Most of the bees hanging from the branch are worker bees, but there are also some drones, and right in the centre of this humming mass, is the queen bee. When they have finally settled and all stragglers have joined, then the humming stops. The whole swarm becomes quiet and hangs like a great golden mass from the branch.

At this point a few bees fly away to scout. They are going to look for a new home for the swarm. But the beekeeper does not want the bees to go and make their home in some old tree, as they might do; instead he covers his head with a veil, and holds a large bag under the hanging swarm. He shakes the branch gently, and thousands of bees fall into the bag. The beekeeper takes the bag to a new hive he has put up, shakes the bees out in front of it, and they quickly move into the new home prepared for them. Once the beekeeper sees the queen bee – she is larger than the others – go through the little opening of the new hive, he does not have to worry any more, for all the other bees come pouring in after her.

In the old hive, before the move, there were about eighty thousand bees and more than half of them move to the new hive with the queen. Now there are about twenty to forty thousand left in the old hive, and there is plenty room for the remaining bees. The bees in the old hive are now without a queen, as the queen has gone with the swarm to the new hive. The old hive cannot be without a queen for long, as bees can't live without a queen. The bees in the old hive have already prepared themselves for this occasion. About two weeks before the swarm day, they made special cells into which they placed eggs, and a new, young queen comes from one of these eggs. So, the old hive has a new queen, and the new hive has the old queen and both are happy.

The queen bee is important because she lays eggs. She does

no work at all other than laying eggs. A queen bee lays about 1500 eggs a day, so she is kept quite busy. The queen bee only ever comes out on the day of swarming; she stays inside the hive and lays eggs the rest of the time. Bee eggs are quite small, only one to two millimetres ($^1/_{16}$ inch). The honey-comb consists of lots of hexagonal wax cells. The worker bees make these cells, and the queen bee moves from cell to cell and lays an egg into each cell.

Where do the bees get the wax? They get it from the nectar of the flowers. Wax comes from the nectar, as well as the honey. If you took a little drop of nectar, however, you would not find any wax in it. There are worker bees that gather the nectar from the flowers, and there are other bees that stay inside the hive and take the nectar from the worker bees that have collected it outside. Most of this nectar is stored in the hive's cells, but some of it is eaten by the bees, and then secreted from little glands in their bodies, coming out as wax. So it is not really eaten in the same sense that we eat things; it is only changed into wax.

There are special bees whose task it is to turn nectar into wax. And there are other bees that take the wax from them, and shape and mould it into the little, regular hexagonal cells that are used to store nectar, and also to store the tiny eggs that the queen lays.

It is wonderful how bees work together, but it is also wonderful how they make exact hexagons from wax, without the slightest mistake, without compass or ruler, and without machinery. This is just one of the many wonders of the beehive.

29. The Beehive

The queen bee lays her eggs in the regular, hexagonal cells, made by the worker bees. The egg lies for three days quietly in its cell, and then it opens and something that does not look the least bit like a bee comes out. It looks like a tiny white caterpillar – it is to the bees what caterpillars are to butterflies – and this bee-caterpillar is called a "larva."

The queen bee does not feed or take care of the little larvae; her only task is to lay eggs. But there are worker bees that have the special task of looking after the larvae. The little things cannot feed themselves; they are fed by the worker bee nurses.

The larva is fed by its nurses for six days, and then it does what the butterfly caterpillar does, it changes into a pupa. For twelve days the pupa lies in its little cell of wax as though dead. After the twelve days have passed, the pupa splits and out comes a little bee. So it is exactly three weeks, twenty-one days, from the time the egg is laid until the bee emerges.

The little bee has three days to get used to the busy life in the hive. And on the third day, the bee becomes a worker itself. No one tells the young bee what it should do, and it is not told how to do it, or when it should do it. Scientists have watched bees carefully and they observed that no older bee stopped its work to spend any time with the young bees; they are all too busy with their own jobs, and yet, the young bee knows what to do. There is no school for bees; they know all they need to know without learning.

First the young bee goes to empty cells and cleans them, as there must not be any dirt or dust in the cells. Bees are very clean creatures and won't allow any dirt in their hive.

After two days of cleaning duties, the young bee becomes a

29. THE BEEHIVE

nurse. It goes to the cells where the honey is stored and takes food to the little larvae. After twelve days of this work the young bee knows, somehow, that it must begin another task, and it becomes a wax maker.

After a time as wax maker, the day comes when the young bee leaves the beehive and flies for the first time. Initially, it only flies around the hive in small circles with its head turned towards the hive, as though afraid it might get lost if it couldn't see its home. It takes a few days practice before the bee is ready to take off for long flights to collect nectar.

Bees have no bags or boxes to carry the sticky liquid nectar with them. Instead they have a second, special stomach – a special bag – in addition to the one they use for eating. The nectar that the bee gathers from the flowers goes into this special stomach until that stomach is full, then the bee flies home to the hive, empties the second stomach, and another bee takes the nectar to store it in cells. The collector bee flies away again to find more nectar.

If one bee discovers an orchard that the other bees of the hive have missed, it has an interesting way of telling its fellow bees that there is a rich treasure of nectar waiting for them. Bees have no language; no voice. (The humming or buzzing we can hear is made by their wings and is no use for telling something to other bees.) Having discovered the orchard, the bee fills its special bag with nectar, flies back to the hive and begins a kind of dance on a honeycomb. It runs in a circle one way, and a curve the other way and waggles its back end, and it seems that from this "dance" the other bees know how to get to the orchard, as they all get quite excited and set off in the right direction, towards the orchard.

There are also other duties for the worker bee. Some worker bees have to stand guard at the entrance to the hive. They touch every incoming bee with the little feelers (antennae) on their heads and recognize if it belongs to the hive or not. There are robber bees and wasps that try to get in and steal the honey, and any intruder is immediately killed with a sharp and poisonous sting. When bees sting, they give up their lives; they loose their sting and they die.

A beehive also needs fresh air inside, just as we need fresh air in a room, but beehives have no windows that can be opened, and they only have a little hole as an entrance. There are always some worker bees on duty near the entrance, and they use their wings like fans, and produce a steady draught of fresh air into the hive.

It is also important to keep a steady temperature in the hive. It must not get too hot in summer, or too cold in winter. The temperature of a beehive is like the temperature of our own bodies, always the same. In fact, the temperature of a beehive is practically the same as the temperature of the human body: 37°C (99°F). When you hold your wet hands in the air they feel cool, or your whole body feels cold when you come out of a bath, because the water evaporates, and it takes away warmth when it evaporates. When we get too hot, the body sweats for the same reason – to cool us down by the evaporation of water.

In the summer, when it gets warmer inside, the bees bring water into the hive and spray the honeycombs and inside walls with it – the bees cool the temperature of the hive down by letting water evaporate inside. They bring water inside to make the hive sweat. We sweat outside the skin; a beehive sweats inside.

In the winter, when it gets too cold, the bees huddle together in the centre of the hive, clinging to one another. The bees on the outside of the cluster wave their feelers, fan their wings, and waggle the back parts of their bodies. In the same way that human beings get warm by moving their arms, so the bees' bodies grow warmer by moving about. This also warms the air inside the hive.

Think of all the different kinds of work that bees do: cleaning, nursing larvae, making wax and wax-cells, collecting nectar, storing nectar in cells, guarding the hive, fanning for fresh air, spraying water in hot weather, and warming the air in cold weather. All this work is carried out by the bees in turn. Every bee knows exactly what kind of job it should do at any given time; there is never a fight about who should do this job or that job; and there is never a job done unwillingly or carelessly. It is truly wonderful that they know how to do all this.

30. The Spirit of the Bee

If you think of all the different work done in a beehive; if you think of the forty thousand bees in the hive and that every one of them knows what to do, how to do it, when to do it; and if you think that each one of the forty thousand bees takes its turn at each of the many tasks, then you will realize that all this would be quite impossible if every single bee had to decide on its own what it should do. It is only possible that all the different jobs are done, and are done at the right time, because there is a mastermind that guides every bee in the hive.

It is this higher mind that guides and controls all the bees, and sees to it that every task is done at the right time, and in the right place; a mind that knows which bee should take a turn at this or that task. You could not say that this higher mind is in the bees, because a bee is, itself, not very clever; one can only say this higher mind hovers over all the bees and makes each bee do what has to be done.

It is different with human beings. We say in the morning verse, "And we to spirit give a dwelling in our soul." Within our own soul, we have the spirit to guide us; it does not hover outside us. And each of us has our own spirit to guide us, not one spirit who controls all of us. This is why we are more aware than bees: our one spirit dwells within us, not outside. Although we are more aware than bees, however, we can learn from bees; we can learn to co-operate, and to work together without envy, jealousy, or greed, as the bees do.

There are three kinds of bee, as you know: the queen, the drone, and the worker bee. The eggs laid by the queen, however, are of two different kinds. From one kind only drones will appear; and from the other kind either workers or a queen will appear. When the little grubs, the larvae, appear from this kind of egg, they are no different from each other. But the larvae are fed by the nurse bees, and the nurses have a choice of two kinds of food to give to the larvae.

On the one hand, there is a very rich sweet milky fluid called royal jelly. The nurse bees make royal jelly from special glands in their heads. On the other hand, there is mixture of pollen and honey. When the bees come back from the flowers, sometimes they carry so much pollen on their legs that they look as though they are wearing yellow trousers. The bees in the hive take this pollen from the field workers and they mix it with honey and store this mixture in cells. This mixture is known as bees' bread.

If the nurses want a larva to become a queen, they feed this larva on nothing but royal jelly. After five or six days, the larva becomes a pupa, and it wraps itself in a little cocoon. From the cocoon a queen bee emerges. It is the royal jelly that makes the larva turn into a queen bee.

This happens only once a year, however, before swarming, when the hive needs a new queen, or when an old queen is about to die. What a hive needs most are worker bees, and larvae that are meant to become workers get royal jelly for the first three days only, after that they are fed on bees' bread. When the larvae become little cocoons and the cocoon opens out, a worker bee emerges.

The hive also needs drones, because the queen would not lay eggs without drones. To make drones, the nurse bees feed the other larvae mostly with bees' bread; they get less royal jelly than all the others. When the cocoons of these larvae open out, the round drones emerge. They have no sting and they do not work. If you recall, the drones are like the pollen in a plant, so it is not surprising that a larva fed mainly on a mixture of pollen and honey becomes a drone.

Let us recall what happens in a flower when the seeds have been touched by pollen, when the seeds have been "pollinated." The petals fall off; the little stems that carried pollen, the stamens, fall off; and only the ovary remains and grows into a fruit with seeds in it. An apple is nothing more than an ovary that has grown very big. The ovary is all that remains and grows, the petals and stamen fall off and die. The drones are like stamens and pollen, so they also have to die.

In a plant, the petals and stamens fall off because the green leaves don't send any more food to them, and the same thing happens in the hive. In autumn, when winter is approaching, the great mind that guides all the bees tells the worker bees not to feed the drones any more. The worker bees, that have given food to the drones for many months, stop feeding them, and the drones die before winter sets in. The drones cannot get any food for themselves, they can't collect nectar, and they can't take nectar from any of the cells in the hive.

It is necessary for the hive that the drones die, for the honey stored in the cells would not be enough to feed the workers and the larvae and the drones through the winter months. The drones die for the good of the hive.

The honey is the bees' store for the winter when no nectar can be found outside. If the beekeeper took all the honey away, and did nothing else, then all the bees would starve and die. The bees make more honey than they can possibly use, so beekeepers are able to take the extra honey without harming the bees. Sometimes a beekeeper does take all the honey and then he puts sugar water into the hive, and although it is not as nourishing as honey, the bees can live on the sugar water until spring when they get nectar from the blossoms again.

The ancient Greeks looked upon the bees with great reverence. They watched the bees flying away from the hive, collecting nectar from the flowers, and returning to the hive, laden with the sweet treasure. The Greeks said: "As the hive is the home of the bee, so the kingdom of heaven is the true home of the human soul." From the kingdom of heaven we come, and we return to it when we die. But we do not return to heaven empty handed. We bring with us all the experiences, all that we have learned in life. We bring all that life has taught us, just as the bee carries nectar back to the hive. And as the bee flies out again for new nectar, so the soul will come to earth again for new experiences, to learn more. That is why the Greeks said: "The soul is like the honey bee."

In some places in ancient Greece, there were temples that were regarded with special awe and reverence by the Greeks. In these temples there were no male priests, but maidens, priestesses. And such a holy maiden was called *melitta*, which means honey-bee. It means a human soul that will return to heaven, rich in wisdom and knowledge, as the bee returns to the hive laden with nectar.

Index

acorn 78f
air 19f
alga 25–27, 29
apple tree 67

bee 51, 53, 99–108
beehive 97f
birch 81–83
blossom 19f, 47–49
briar-rose 65
brussel-sprout 70
bulb 46, 57f
buttercup 47
butterfly 51, 53–56, 73

cabbage 69–71
–, wild 71
calyx 46–48, 64f
camellia 87
carnation 44, 46
caterpillar 54f
cauliflower 70
cereal 91–94
chrysalis 54–56
cocunut palm 85f
coffee 90
conifer 36–38
copra 86
cotton 90
cotyledons 60–62
crocus 45
crown of a tree 81

daffodil 45
dandelion 18–20
Demerara sugar 89
dicotyledons 62

drone 98, 100, 105–7
dwarf-birch 83

earth 19f, 39f
egg 54f

ferns 33–35
fir 37
flowering plant 41–43
fructose 88
fruit 19f, 51f
fruit trees 67
fungus 21–24, 29

gall wasp 79
gall-apple 79f
glucose 88
grass 91–94
green leaf 19f

hollyhock 90
honey 101
horsetail 34f

iris 45

jay 79

kohlrabi 70

larch 37
larva (of bee) 102
leaves 95f
lichen 28–30
light 19
lily 45

maize 93
molasses 89
monocotyledons 62
moon 61
moss 31f

narcissus 45
nectar 51
nectar 103
nettle 72–74

oak 77–80
onion 58f
ovary 48–50, 73, 96f, 107

palm tree 84–86
pansy 60
petal 48
pine 37
pistil 48f, 73
pollen 48–52, 97, 106f

queen bee 98, 100–102, 105f

red admiral 73
resin 38
rice 93
root 20
rose 46, 60, 63–68
– family 66–68
–, wild 65
royal jelly 106

sap 82
seed 60
sepal 48, 64f
shrub 64
snowdrop 45
soil 77
spore 22f
stamen 48–50, 73
stars 42
sugar 88
– beet 89
– cane 88
sunflower 44

tea 87f
tree 39f
truffle 23
tulip 45, 57–60, 70

veins, parallel 45, 60
veins, reticulate 45
violet 44, 46

warmth (from earth) 19f
warmth (of sun) 19f
water 20, 31f
water lily 91
wax 101, 103
wild cabbage 70
wild rose 65
worker bee 98, 100f, 105f

Other Waldorf Education Resources by Charles Kovacs

Class 4 (age 9–10)
    Norse Mythology

Classes 4 and 5 (age 9–11)
    The Human Being and the Animal World

Classes 5 and 6 (age 10–12)
    Ancient Greece
    Botany

Class 6 (age 11–12)
    Ancient Rome

Classes 6 and 7 (age 11–13)
    Geology and Astronomy

Class 7 (age 12–13)
    The Age of Discovery

Classes 7 and 8 (age 12–14)
    Muscles and Bones

Class 8 (age 13–14)
    The Age of Revolution

Class 11 (age 16–17)
    Parsifal and the Search for the Grail

General interest
    The Spiritual Background to Christian Festivals

For news on all our **latest books**, and to receive **exclusive discounts**, **join** our mailing list at:

florisbooks.co.uk

Plus subscribers get a FREE book with every online order!

We will never pass your details to anyone else.